# The International
# MUSSEL
# WATCH

# The International MUSSEL WATCH

Report of a Workshop Sponsored by the
Environmental Studies Board
Commission on Natural Resources
National Research Council

NATIONAL ACADEMY OF SCIENCES
Washington, D.C.    1980

This workshop was supported by the U.S. Environmental Protection Agency, Contract No. 68-01-2430.


Library of Congress Catalog Card Number 80-80896

International Standard Book Number 0-309-03040-4

Available from

Office of Publications
National Academy of Sciences
2101 Constitution Avenue, N.W.
Washington, D.C.  20418

Printed in the United States of America

ENVIRONMENTAL STUDIES BOARD

CONTENTS

LIST OF FIGURES

LIST OF TABLES

PREFACE

The integrity of the world's coastal waters is
jeopardized by the deliberate and inadvertent entries of
society's discards.  Many substances introduced by mankind
are toxic to marine organisms, thus impinging upon the
health of ocean communities or restricting the human
consumption of fish and shellfish.  Other substances can
interfere with the use of coastal zones for recreation or
transportation.  Those responsible for the management of
coastal waters need a continuous assessment of the extent of
contamination so that remedial actions can be taken to
prevent or minimize loss of our various and unique marine
resources.

The scientific community strives to understand the
health of coastal waters, but there are many obstacles, both
scientific and economic, that limit its activities.  The
number of potential contaminants is great and the pollutants
themselves are often present in the waters in concentrations
so minute that accurate analyses tax the abilities of even
the most expert chemists.  In addition, monetary resources
are limited with which to carry out analyses frequently
enough and with sufficient geographical coverage to improve
our understanding of pollution problems.  As a consequence,
the use of sentinel organisms--creatures of the sea that
concentrate within their bodies a number of polluting
substances--offers great advantages to both scientists and
environmental managers.

Out of such considerations, "Mussel Watch" was born.  It
is a strategy of using bivalves, like mussels, oysters, and
clams, as recorders of environmental levels of pollutants.
The technique was used effectively in northern Europe and
the United Staes from the late 1960s through the 1970s, and
it has since been adopted by an ever increasing number of
nations.

Early in 1978, it became evident that a great deal of
information of value to those planning or already involved

in a "Mussel Watch" was available--often in unpublished
form--from scientists studying the natural history of
bivalves or their use as sentinel organisms.  The insight
led to plans for an International Workshop to bring such
scientists together to assess strengths and weaknesses of
the Mussel Watch concept, and to propose research whose
results could enhance the concept's usefulness.

The Workshop, held in Barcelona, Spain, December 4 to 7,
1978, was sponsored by the National Academy of Sciences of
the United States through its Environmental Studies Board.
The 57 participants in the Workshop, representative of 20
countries, were divided into 6 panels.  Four panels were
dedicated to pollutants amenable to study using bivalves:
fossil fuels and their combustion products, trace metals,
halogenated hydrocarbons, and artificially produced
radionuclides.  A fifth group assessed current techniques
for using measures of disturbances to the health of mussels
from pollutants to seek out unacceptable pollutant levels.
The sixth group surveyed tactics for carrying out a
successful mussel watch program.

Each panel considered its subject in a different way
and, as a result, each chapter in the report has its own
flavor and mood.  The panel on radioactivity, for example,
used the results of the first year of the U.S. Mussel Watch
to evaluate recent procedures and propose additional
research; the fossil fuels panel surveyed results from a
broader geographical base.  All the panels recognized and
cited deficiencies in current methodologies.

The report presents the findings of the 6 panels.  Aside
from arriving at general agreement that the concept of a
global mussel watch program has merit, the Workshop did not
attempt to develop a set of comprehensive guidelines or
recommendations for its implementation.  In some instances,
however, individual panels did arrive at conclusions or
suggestions for carrying out particular parts of the overall
program, and these are included in the panel reports.  For
example, some of the panels asked the question, "How many
mussels are necessary to provide an adequate representation
of a population for chemical assay?"  The number 25 appeared
to be generally accepted.  The conclusions and
recommendations of the panels have been left in the
individual panel reports and not brought together in a
single chapter since the participants never met for the
purpose of consolidating recommendations.  In addition, to
dissociate the recommendations from the context of the
discussions in which they arose would have diminished the
effectiveness of the report.

The report provides a "state-of-the-art" evaluation of the use of bivalves in marine monitoring. Its primary audience will be those investigators concerned with coastal zone management, but it will also be of value to researchers in international programs examining near-coastal pollution and effects of ocean dumping. Clearly, as more extensive use of the techniques described here occurs, the accumulating experience will provide guidance for more effective monitoring programs. A second International Mussel Watch Workshop has been proposed, and it is hoped that the scientists gathered there will find the present volume a useful base on which to build their new findings and recommendations.

The participants in the Workshop would like to express deep appreciation to the hosts, Antonio Ballester Nolla of the Instituto de Investigaciones Pesqueras and Enrique Balcells of the Spanish National Committee to Scope. Thanks are also due to Janis Horwitz, Raphael Kasper, and William Robertson IV of the U.S. National Academy of Sciences and to Elaine Blanc, of the SCOPE Secretariat, without whose efforts the Workshop and this volume would not have been possible.

EDWARD D. GOLDBERG, Chairman
International Mussel Watch Workshop

# LIST OF ABBREVIATIONS

AOAC    —    Association of Official Analytical Chemists
BaP     —    benzo(a)pyrene
BCI     —    body component index
DDE     —    dichlorobischlorophenylethylene
DMSO    —    dimethyl sulfoxide
DDT     —    dichlorodiphenyltrichloroethane
EC      —    energy charge
EPA     —    Environmental Protection Agency
FAO     —    Food and Agriculture Organization
GLC     —    gas liquid chromatography
HC(s)   —    hydrocarbon(s)
HPLC    —    high-pressure liquid chromatography
IAEA    —    International Atomic Energy Agency
ICES    —    International Council for Exploration of the Sea
IDOE    —    International Decade of Ocean Exploration
NBS     —    National Bureau of Standards
NOAA    —    National Oceanic and Atmospheric Administration
NRC     —    National Research Council
ODS     —    octadecylsilic
OECD    —    Organization for Economic Cooperation and Development
PAH(s)  —    polynuclear aromatic hydrocarbons
PCB(s)  —    polychlorinated biphenyls
SRM     —    standard reference material
TDE     —    tetrachlorodiphenylethane
TLC     —    thin-layer chromatography
UV      —    ultraviolet
UCM     —    unresolved complex mixture
UNEP    —    United Nations Environment Program

# OVERVIEW

Some alterations of the environment caused by human activities, such as air pollution from large urban centers and changes in the landscape from surface mining, are obvious. But the complex effects of other human activities, like the long-term interactions of components of the global environment and long-distance transport of pollutants, are less obvious and, for that reason, less tractable. Alterations of the chemical composition of the atmosphere, fresh waters, coastal waters, and oceans through the introduction of anthropogenic and natural substances and their subsequent effects on living organisms bring into play large-scale interrelationships of environmental systems and the possibility of effects ranging far beyond the original sites of human activity. Unfortunately, only a partial picture exists of how extensive those effects may be, and the needs for filling in that picture are increasing. The impacts on the coastal marine environment can jeopardize resources of great economic importance to many nations.

There are compelling reasons to initiate a systematic and continuing program for monitoring contaminants that are introduced into the coastal marine environment. The coast is where the atmosphere, land, and water meet. It is a dynamic place characterized by environmental extremes and rapid interactions. It is in close proximity to human activities and highly susceptible to them. These attributes combine to make the world's coastal environment a logical place to initiate a global monitoring effort.

Monitoring strategies may be as diverse as the problems or needs to which the programs are addressed. Monitoring will be successful only to the degree that it responds to objectives that must be clearly defined before any monitoring program is initiated. The objectives of a global mussel watch or, indeed, of any coastal monitoring system should be:

1. to advance the state of knowledge and understanding of environmental processes;
2. to support the processes of environmental regulation, standard setting, and enforcement;
3. to determine and assess the level of contamination of coastal areas and warn of potentially dangerous conditions;
4. to develop methods and instrumentation; and
5. to train scientists.

A global mussel watch to meet these objectives must be designed by those scientists who will implement the program as well as by those responsible for the management of the coastal environment. Considerations that they must weigh are outlined in the sections of this report dealing with classes of identified pollutants.

Contaminants introduced into coastal waters pose possible hazards to human health through the consumption of seafood, through direct exposure, or through damage to ecosystems. The results are potential losses of marine resources or restricted uses of them. The effects of unpredictable alterations of ecosystems are unknown and troubling.

Although these concerns are well recognized and commonly discussed, no action has been taken to establish compatible monitoring programs on a global basis for contaminants already identified as potentially presenting problems. The contaminants are petroleum components, halogenated hydrocarbons, heavy metals, and synthetic radionuclides. There have, however, been effective efforts toward monitoring local and regional coastal contamination in the past and we have attempted to build upon those experiences. We have tried to restrict our recommendations to activity that could form the core of a useful and productive program of world environmental cooperation.

We share the conviction of a growing number of scientists that large and encompassing studies are not appropriate. Plans for environmental monitoring should consider what portion of the world's finite resources of scientists and funds it is appropriate to commit to the program. Well-defined marine activities are needed now to help us understand the interactive global environment and to produce results that will be of value to the world community. The immediate aim is to concentrate upon those few activities that promise substantial returns from limited support. A global mussel watch has the potential for meeting these goals.

Most of the pollutants identified as potentially significant environmental problems have seawater concentrations in the range from $10^{-15}$ grams per gram (g/g) to $10^{-12}$ g/g. Although methods for the measurement of seawater pollutants at these low concentrations are now being developed, they are not as yet in routine use. Bivalves are suitable animals for such measurements because of their ability to concentrate some substances found in seawater, thus circumventing the necessity to obtain, transport, and process large volumes of seawater that must be kept free from contamination throughout the extraction and analytical steps. Bivalves can play a key role in the analytical scheme by providing an initial enrichment step, an approach used previously by Goldberg and others (1978) and Holden (1973). There are a number of unresolved problems in the use of bivalves as sentinel organisms that have been revealed by earlier work and that were pointed out at the workshop. For example, some forms of pollution cannot be monitored by bivalves; they include biostimulants such as nitrate and phosphate, and perhaps low-molecular-weight halocarbons such as chloroform and carbon tetrachloride. While acknowledging the difficulties and the need for continuing research, we argue that enough is known now to pursue a productive program of global coastal pollution monitoring using the mussel watch strategy.

An essential phase complementary to monitoring is assessment of results. Initial assessments should be made by actively involved scientists who understand the limitations of the data.

The oversight and management of the activities of a global mussel watch program should rest with a guidance committee representing the world community of scientists, both within and outside government. The guidance committee should determine how the global mussel watch will be designed, implemented, coordinated, and assessed. These activities will require scientists with firsthand knowledge of the monitoring program and of its uses.

The program should be carried out by scientists and nations cooperating throughout the world and should be dependent upon the support of the nations involved. Therefore, it is important that the advisory committee, as well as the working teams of scientists, be international in composition.

The global mussel watch has two general aims: (1) to produce practical information on contemporary problems, such as the extent of contamination of coastal ecosystems and food resources; and (2) to advance the basic scientific

understanding of the world's coastal systems and their
dynamics by providing intercomparable global data on the
abundance of anthropogenic contaminants. Implicit in what
is proposed is a conceptual model of the physical,
biological, and social systems affected. It is impossible
to construct any monitoring system in the absence of such a
concept, and its development must be a primary goal of the
program.

Control of the quality of the data to be used will be
critical to the success of the system. All data should have
associated with them measures of accuracy and precision.
The preparation of standards and intercalibration exercises
involving all laboratories is essential. This will be time-
consuming and relatively expensive. Whatever the range of
costs, the data produced will be of little value without
intercalibration (Farrington and others 1973, Holden 1973,
Farrington and others 1976). Laboratories might be selected
to serve as centers responsible for intercalibration for a
limited time and on a rotating basis. Intercalibrations
should be run periodically, as appropriate to the
observations in question. The results of intercalibration
exercises should be made promptly available to all
participating laboratories.

To ensure that the data obtained by mussel watch provide
an unbiased estimate of the variations of the natural
system, the sampling program must be statistically sound.
We emphasize this as another key factor in the success of
the effort. The design must include determination of both
the component of variance that derives from differences
among individual samples and the component of variance that
derives from the precision of the measurements. If good
estimates of the variation among samples are available,
samples may be pooled, eliminating the necessity for a
series of costly, time-consuming analyses.

There is a need for rapid dissemination of monitoring
data to research workers and an equally essential
requirement for feedback from the research laboratory to the
monitoring program.

Data storage and retrieval arrangements will have a
significant impact on the success of the effort. The data
files should be open and accessible to the scientific
community. Time requirements for reporting monitoring data
and introducing them to the open files should be established
and kept as short as possible. Measures of quality, such as
the intercalibration results from the originating
laboratory, should be included.

Another important principle is the long-term continuity
of the data files.  The monitoring record may be continuous
over decades, and it will become more valuable as the data
files expand.  Intercalibration laboratories should be
responsible for keeping abreast of analytical advances and,
in concert with the advisory committee, for introducing new
techniques in such a manner that continuity of data is
maintained.

It is inevitable that information will become available
that will call into question the data or their
interpretation.  In anticipation of this state of affairs it
will be valuable to preserve samples for reexamination in
clearing up uncertainties.  Wherever practicable, an
appropriate library of samples should be maintained and
catalogued.  Methods of preservation and use of samples
should be controlled by the advisory committee, and
repositories for the samples should be maintained in
permanent research institutions.

Any country can, in principle, develop the capacity for
making many of the measurements reported here in a
relatively short time.  A commitment to conscientious and
careful work is all that is required.  However, measurements
of such substances as dioxins or curium isotopes require a
level of expertise that may not be readily transferable.
For these measurements, samples should be sent to
established laboratories for analysis.  Continual contact
among laboratories participating in the mussel watch will
provide a basis for establishing and maintaining the quality
of the data.  Once their capabilities are developed, the
laboratories will become centers for dealing with local and
regional problems, as well as components in a system of
international exchange among scientists.

The International Mussel Watch Workshop, held in
Barcelona, Spain, December 4 to 7, 1978, was convened by the
U.S. National Research Council (NRC) and chaired by Edward
D. Goldberg of the Scripps Institution of Oceanography.  The
workshop had five major goals:

1.  to assess the use of bivalves in determining
environmental concentrations of chemical pollutants and
pathogens;
2.  to evaluate current data with respect to
distinguishing natural from pollutant concentrations for
heavy metals and hydrocarbons and determining changes in the
amount of pollution as a function of time;

3. to formulate effective biological monitoring strategies to complement the measurement of pollutant levels;

4. to appraise existing techniques for analysis and to propose acceptable methods of carrying out interlaboratory comparisons and standards for all collectives of pollutants; and

5. to consider the expansion of the U.S. Mussel Watch Program to a worldwide basis as a continuous monitor of the health of the coastal waters.

The workshop was divided into panels, each dealing with a specific topic. The topics were: organochlorides, trace metals, petroleum hydrocarbons, transuranics, mussel health, and monitoring strategies. The following chapters contain reports of the discussions of each of these panels.

## REFERENCES

Farrington, J.W., J.M. Teal, T.G. Quinn, T. Wade, and K. Burns (1973) Intercalibration of analyses of recently biosynthesized hydrocarbons and petroleum hydrocarbons in marine lipids. Bull. Environ. Contam. Toxicol. 10:129-136.

Farrington, J.W., J.M. Teal, G.C. Medeiros, K. Burris, E.A. Robinson, Jr., J.G. Quinn, and T.L. Wade (1976) Intercalibration of gas chromatographic analyses for hydrocarbons in tissues and extracts of marine organisms. Anal. Chem. 48:1711-1715.

Goldberg, E.D., V.T. Bowen, J.W. Farrington, G. Harvey, J.H. Martin, P.L. Parker, R.W. Risebrough, W. Robertson, E. Schneider, and E. Gamble (1978) The mussel watch. Environ. Conserv. 5(2):101-25.

Holden, A.U. (1973) International cooperative study of organochlorine and mercury residues in wildlife, 1969-1971. Pestic. Monit. J. 7:37-52.

# 2

FOSSIL FUELS

JOHN W. FARRINGTON (Chairman), Woods Hole Oceanographic
    Institute, Woods Hole, Massachusetts
JUAN ALBAIGES, International Congress on Analytical
    Techniques in Environmental Chemistry, Barcelona, Spain
KATHRYN A. BURNS, The Ministry for Conservation, East
    Melbourne, Australia
BRUCE P. DUNN, Cancer Research Center, The University of
    British Columbia, Vancouver, British Columbia, Canada
PETER EATON, Environmental Protection Service, Halifax, Nova
    Scotia, Canada
JOHN L. LASETER, University of New Orleans, New Orleans,
    Louisiana
PATRICK L. PARKER, University of Texas, Port Aransas, Texas
STEVEN WISE, National Bureau of Standards, Washington, D.C.

INTRODUCTION

Fossil fuel compounds are defined as (a) hydrocarbons
and other organic compounds contained in petroleum, gas,
coal, and oil shale; products of coal and oil shale
processing; products of petroleum, gas, and coal combustion;
and (b) organic compounds that result from chemical,
photochemical, microbial, or other metabolic transformations
of the preceding materials. Examples of classes of
compounds and individual compounds are given in Figures 2.1
through 2.3.

Because of the accelerated demand for energy during the
past half century, pollution of the marine environment by
fossil fuels and their reaction products has been
increasing. Significant progress in determining the
sources, fates, and effects of fossil fuel compounds in the
marine environment has resulted from the decade of research
from 1968 to 1978 (NOAA 1978). Before 1968, investigations
focused mainly on visible slicks, the aesthetic problem of

7

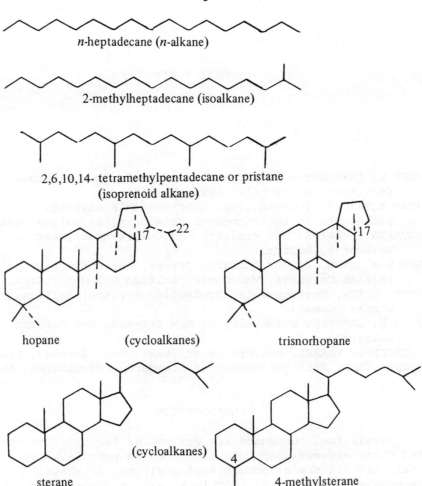

FIGURE 2.1 Structure of alkanes and cycloalkanes (personal communication, J. W. Farrington, Woods Hole Oceanographic Institution, 1979).

tar on beaches, and the acute effects of oil spill on birds. By the early 1970s, it was recognized that the lack of a visible slick did not mean that oil had been removed from the environment (NRC 1975). After slicks had disappeared, measurable amounts of oil in water, biota, and sediments of coastal ecosystems could persist for periods ranging from days to years.

Careful consideration of possible sources of fossil fuel compounds and a few key measurements have led to the realization that oil spilled as a result of tanker and

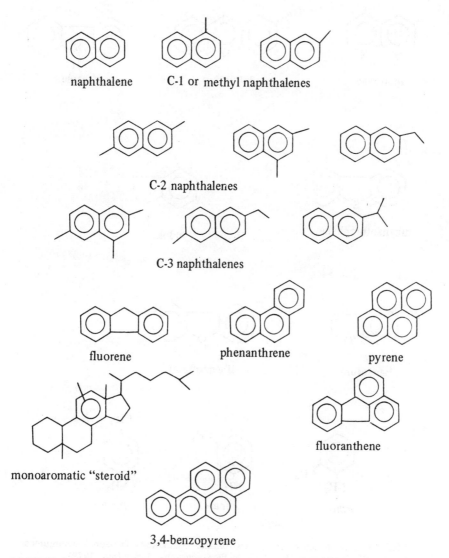

naphthalene          C-1 or methyl naphthalenes

C-2 naphthalenes

C-3 naphthalenes

fluorene          phenanthrene          pyrene

fluoranthene

monoaromatic "steroid"

3,4-benzopyrene

FIGURE 2.2   Structure of aromatic hydrocarbons (personal communication, J.W. Farrington, Woods Hole Oceanographic Institution, 1979).

offshore drilling and production accidents represents only a small part of the total input of fossil fuel hydrocarbons to the marine environment.  For example, sewage and industrial effluents; river and urban runoff; atmospheric rainout and fallout; and routine tanker operations, such as cleaning of

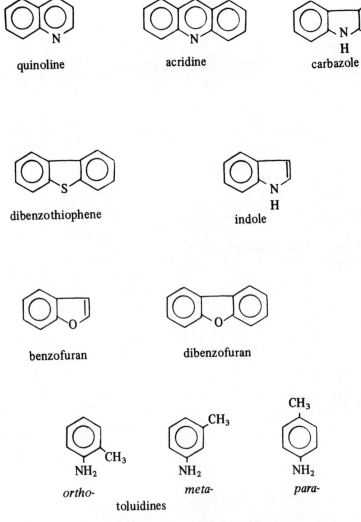

FIGURE 2.3 Structure of heteroaromatic compounds (personal communication, J. W. Farrington, Woods Hole Oceanographic Institution, 1979).

tanks, have all been identified as real or potentially significant sources of such compounds.

Although world petroleum consumption is expected to decrease substantially between 1990 and 2010, the use of coal and oil shale as energy sources and chemical feedstock is expected to increase for a century or more. Coal gasification and liquefaction, shale processing, transport, and combustion release many of the same fossil fuel compounds into the environment as do oil combustion, oil spills, and effluent discharge. The prevention of adverse impacts on living organisms and the environment will require knowledge of the biogeochemistry of fossil fuel compounds and their reactive products.

Bivalves have been used in local, national, and regional studies of fossil fuel pollution--both chronic and acute (Kidder 1977). To date, analyses of bivalves have provided useful information about fossil fuels in the environment and will continue to do so in the future. However, bivalves are only a small part of the marine environment and we caution that while they have been and will continue to be valuable as sentinel organisms, we need to know more about the general biogeochemistry of fossil fuel compounds in the environment if we are to interpret the information available from analyses of bivalves.

We set forth here an assessment of our current knowledge of the concentrations and sources of fossil fuel compounds in bivalves, temporal and spatial distributions, and processes governing these distributions. We also briefly describe principles and methods of measurements.

Of the many organic compounds released to the environment, some are known or suspected to pose threats to public health or the integrity of ecosystems (NRC 1976). The chlorinated hydrocarbons discussed in Chapter 4 of this report and the fossil fuel compounds are among these compounds. Recently, the U.S. Environmental Protection Agency (EPA) issued a list of priority pollutants that contained several of the chlorinated hydrocarbons and fossil fuel compounds (Table 2.1). Few, if any, measurements in marine organisms have been reported for many of the organic compounds. Yet there is a need to know their distributions and biogeochemistry in coastal ecosystems to assess the magnitude and location of any threats to living organisms.

## FOSSIL FUEL COMPOUNDS IN BIVALVES

We present here results from representative programs of analyses of fossil fuel compounds in bivalves. Since most of the laboratories involved have not intercalibrated their results in a rigorous manner, strict comparison of data is not possible at this time. Nevertheless, conclusions based on the results within each program can be compared.

TABLE 2.1   Organic Compounds in the U.S. Environmental Protection
Agency's List of 129 Unambiguous Priority Pollutants

*Volatiles—33*

Acrolein
Acrylonitrile
Benzene
*Bis*(chloromethyl)ether
*Bis*(2-chloroethyl)ether
*Bis*(2-chloroethoxy)methane
Bromoform
Carbon tetrachloride
Chlorobenzene
Chlorodibromomethane
Chloroethane
Chloroform
Dichlorobromomethane
Dichlorofluoromethane
Ethyl benzene
Methyl bromide
Methyl chloride
Methylene chloride
Tetrachloroethylene
Toluene
Trichloroethylene
Trichlorofluoromethane
Vinyl chloride
1,1-Dichloroethane
1,1-Dichloroethylene
1,1,1-Trichloroethane
1,1,2-Trichloroethane
1,1,2,2-Tetrachloroethane
1,2-Dichloroethane
1,2-Dichloropropane
1,2-*Trans*-dichloroethylene
1,3-Dichloropropene
2-Chloroethyl vinyl ether

*Pesticides and PCBs—28*

Aldrin
α-BHC
β-BHC        Hexachlorobenzenes
Δ-BHC
γ-BHC        Lindane
Chlordane
Dieldrin
Endosulfan sulfate
Endosulfan, β

*Pesticides and PCBs—28*

Endosulfan, α
Endrin
Endrin aldehyde
Heptachlor, epoxide
Heptachlor
Hexachlorocyclopentadiene
Hexachlorobutadiene
Hexachlorobenzene
PCB 1221
PCB 1232
PCB 1242
PCB 1248
PCB 1254
PCB 1260
PCB 1016
Toxaphene
4,4'-DDD
4,4'-DDE
4,4'-DDT

*Other Neutrals—10*

Bis-2-chloroisopropyl ether
Hexachloroethane
1,2-Dichlorobenzene
1,2,4-Trichlorobenzene
1,3-Dichlorobenzene
1,4-Dichlorobenzene
2-Chloronaphthalene
4-Bromophenyl phenyl ether
4-Chlorophenyl phenyl ether
Isophorone

*Phenols—11*

p-Chloro-m-cresol
Pentachlorophenol
Phenol
2-Chlorophenol
2-Nitrophenol
2,4-Dichlorophenol
2,4-Dimethylphenol
2,4-Dinitrophenol
2,4,6-Trichlorophenol
4-Nitrophenol
4,6-Dinitro-o-cresol

## TABLE 2.1 (Continued)

| N-Containers-4 | Phthalates-6 |
|---|---|
| | |

*N-Containers-4*

1,2-Diphenyl hydrazine
Nitrobenzene
2,4-Dinitrotoluene
2,6-Dinitrotoluene

*Hetero-Carcinogens-6*

*N*-nitroso-di-*n*-propyl amine
*N*-nitroso-diphenyl amine
*N*-nitroso-dimethyl amine
2,3,7,8-Tetrachlorodibenzo(p)dioxin
Benzidine
3,3'-Dichlorobenzidine

*Phthalates-6*

*Bis*(2-ethyl hexyl) phthalate
Butyl benzyl phthalate
Di-*n*-butyl phthalate
Diethyl phthalate

*Phthalates-6*

Dimethyl phthalate
Di-*n*-octyl phthalate

*Hydrocarbons-16*

Acenaphthylene
Acenaphthene
Anthracene
Benz(e)acephenanthrylene
Benzo(k)fluoranthene
Benzo(a)pyrene
Chrysene
Fluoranthene
Fluorene
Indeno (1,2,3-cd)pyrene
Naphthalene
Phenanthrene
Pyrene
Benzo(ghi)perylene (1,12 benzoperylene)
Benzo(ghi)anthracene (1,2 benzanthracene)
1,2,5,6-Dibenzanthracene

SOURCE: Reprinted with permission from Environ. Sci. Technol., vol. 13, L. H. Keith and W. A. Telliard, ES&T special report priority pollutants: I. A perspective view, copyright 1979, American Chemical Society.

## Westernport and Port Phillip Bay, Australia[1]

Mytilus edulis collected from locations adjacent to known or suspected sources of fossil fuel compound input and from fairly clean remote sites were analyzed for hydrocarbons. A total of 20 intertidal and offshore sites were sampled at intervals of approximately 6 months in Westernport Bay during 1975-1976. Uncontaminated mussels from a clean area were transferred to other areas of Westernport Bay for monitoring purposes.

Mussels from areas influenced by a refinery outfall showed high persistent values for petroleum contamination of 600 to 1,200 parts per million (ppm) dry weight measured as an unresolved complex mixture signal in gas chromatograms. Progressively lower concentrations were found in mussels sampled at sites at increasing distances from input areas until no petroleum hydrocarbons were found. Bay areas influenced by chronic oil pollution were mapped to show levels of petroleum hydrocarbons in tissues (Figure 2.4).

# WESTERNPORT BAY

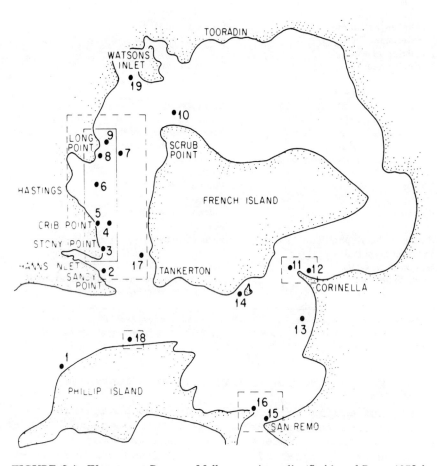

FIGURE 2.4   Westernport Bay near Melbourne, Australia (Smith and Burns 1978a).

A similar program was undertaken at 56 stations in Port Phillip Bay.  The two most heavily contaminated areas are the Corio and Hobson segments (see Figure 2.5).  Oil contaminants found in mussels were assumed to be from mixed sources, such as ships, industrial discharges, and urban street runoff.  Refinery effluent is a major source of contamination in the Corio Segment, while discharges from

# PORT PHILLIP BAY

FIGURE 2.5  Port Phillip Bay near Melbourne, Australia (see text for further information) (Smith and Burns 1978b).

ships into the Yama River system constitute major inputs into the Hobson Segment.

Mussels in both Port Phillip Bay and Westernport Bay showed differences of at least two orders of magnitude in concentrations of petroleum hydrocarbons; those near known or suspected sources of input contained the higher concentrations.

Benzo(a)pyrene in Bivalves from
U.S. and Canadian Coastal Areas[2]

Benzo(a)pyrene (BaP) concentrations in 45 bivalve
samples along the Oregon coast of the United States have
been measured by Mix and others (1977). Concentrations
ranged from about 0.1 parts per billion (ppb) dry weight to
30 ppb dry weight. Bivalves from industrialized bays
contain elevated concentrations of BaP compared to bivalves
from nonindustrialized bays.

Dunn and Stich (1975) report that bivalves at stations
remote from harbor areas contained very small amounts of BaP
ranging down from 0.2 ppb wet weight. Mussels collected
from the Vancouver, B.C. outer harbor contained higher
concentrations, averaging 2.0 ± 3 ppb wet weight, and
mussels from the poorly flushed inner harbor contained 42 ±
6 ppb wet weight.

Dunn and Young (1976) reported analyses of BaP in
mussels from 25 stations along the southern California
coast. The samples were taken mostly from areas on the open
coast. The results indicated that even in heavily populated
coastal areas, mussels taken from locations removed from
local sources of pollution contained very low concentrations
of BaP, about 0.1 ppb wet weight.

Eaton and Zitko (1979) analyzed several samples of
bivalves at 70 locations along the Atlantic coast of Canada
for several polynuclear aromatic hydrocarbons (PAHs).
Concentrations in shellfish were generally higher close to
creosoted wood structures such as wharves or bridges and
decreased with increasing distance from the structures.
Contamination was considerable within 100 meters of the
structures, with geometric means of 800 ppb dry weight for
some individual compounds such as fluoranthane.

## U.S. Mussel Watch Data

The locations of stations in the U.S. Mussel Watch
Program are given in Goldberg and others (1978) and shown in
Figure 2.6. Elevated concentrations of fossil fuel
compounds were found in mussels and oysters from a number of
areas (Tables 2.2 and 2.3). Concentrations of individual
PAHs at some stations near harbor areas were at least two
orders of magnitude higher than those in areas remote from
industrial activities along the coast; see, for example, the
data for Boston Harbor (Table 2.4).

17

In some cases, the distribution of alkylated aromatic hydrocarbons and parent hydrocarbons suggest that the primary source of most of the fossil fuel input was crude oil or fuel oil.  In several other samples, the alkylated aromatic hydrocarbon/parent hydrocarbon ratios indicate that a good portion of the aromatics were from pyrolytic sources.

The preceding are a few examples of the use of bivalves in monitoring specific harbor areas and in surveying coastal areas for fossil fuel compounds.

A general consensus from these studies is that elevated concentrations as much as two orders of magnitude above background in remote areas are found near known or suspected sources of inputs of fossil fuel compounds.  There is sufficient variety in the types of areas monitored to conclude that bivalves are useful sentinels for general survey and monitoring programs for fossil fuel compounds in coastal areas.  Detailed analyses often make possible identification of probable major sources for the fossil fuel compounds, for example, the pattern of parent and alkylated aromatic hydrocarbon distributions.  The exact relationship between concentrations in bivalves and concentrations in surrounding ecosystems is not known, but our understanding is increasing.  More detailed analyses should result in more definitive identification of sources of fossil fuel compounds.

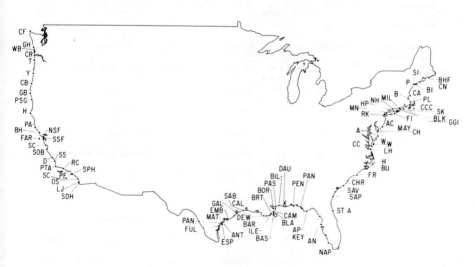

FIGURE 2.6  U.S. Mussel Watch stations (Goldberg and others 1978) (*see overleaf*).

## Station Locations

| Location | Station Initials[a] and Date[b] | Fix | Species | Water Temp. | Description of Site |
|---|---|---|---|---|---|
| Bodega Head | BH 760324 | 38° 14'N 123° 03.8'W | M. californianus | | Horizontal tidal ledge, upper range |
| Pt. La Jolla | LJ 760420 | 32° 51'N 118° 16.6'W | M. californianus | 15°C | Horizontal tidal ledge, upper middle range |
| San Diego Harbor | SDH 760425 | 32° 40.9'N 117° 14.2'W | M. edulis (1,2,3,5) M. californianus (4)[c] | 18°C | 1, 2, from concrete bridge standards 3, N. tip of Harbor Island 4, Point Loma, just inside harbor on granite breakwater 5, North Island on abandoned berths • |
| Point Fermin | PF 760506 | 33° 42.9'N 118° 19'W | M. californianus | 16°C | Horizontal tidal ledge, upper reaches of mussel range |
| Rincon Cliffs | RC 760507 | 34° 21.3'N 119° 26.5'W | M. californianus | 15°C | Breakwater rocks, base of pier |
| Point Arguello | Pt A. 760508 | 34° 34.7'N 120° 38.8'W | M. californianus | 12°C | Tip of Arguello, N. side |
| San Pedro Harbor | SPH 760516 | 33° 42.8'N 118° 16.4'W | M. edulis | | 1, Rock and concrete landfill 2, Large granite rocks, base of pier, at entrance to Middle Harbor 3, Concrete pier pilings 4, Concrete pilings, Terminal Island side of Main Channel 5, Terminal Island side of Main Channel, below bridge and near berth |
| Oceanside jetty | OS 750623 | 33° 11.6'N 118° 23.4'W | M. californianus | 17°C | Granite beach jetty |
| Santa Catalina Is. | SC 760604 | 33° 27.1'N 118° 29.2'W | M. californianus | | |
| San Simeon | SS 760614 | 35° 37.2'N 121° 09.2'W | M. californianus | 13°C | Horizontal tidal ledge |
| Diablo Canyon | D 760616 | 35° 12.1'N 120° 50.6'W | M. californianus | 13°C | Horizontal rock surface, upper range of mussels |
| Santa Cruz | SC 760622 | 37° 57.1'N 122° 04.9'W | M. californianus | 12°C | Horizontal tidal ledge, eroding sedimentary rock |
| Soberanas Pt. | SOB 760623 | 36° 27'N 122° 55.7'W | M. californianus | 11°C | Granite semi-horizontal ledge, upper-mid range |
| N. San Francisco (11 stations) | NSF 760701 (1) | 38° 0.38'N 122° 13.5'W | M. edulis | 17°C | Concrete pilings |
| | (2) | 38° 3.4'N 122° 15.8'W | M. edulis | 19°C | Concrete pier pilings |
| | (3) | 38° 3.3'N 122° 18.3'W | M. edulis | 19°C | Metal buoy |
| | (4) | 38° 2.4'N 122° 20.9'W | M. edulis | 19°C | Metal buoy |
| | (5) | 38° 0.7'N 122° 24.1'W | M. edulis | 20°C | Metal buoy |
| | (6) | 37° 57.8'N 122° 25.9'W | M. edulis | 18°C | Creosote wood pier pilings |
| | (7) | 37° 56.1'N 122° 26.5'W | M. edulis | 19°C | Creosote wood pier pilings |
| | (8) | 37° 51.8'N 122° 23.7'W | M. edulis | 18°C | Metal buoy |
| | (9) | 37° 49.9'N 122° 28.7'W | M. edulis | 16°C | Creosote pylon |
| | (10) | 37° 50.9'N 122° 21.5'W | M. edulis | 18°C | End of pier, 1 m above water level |
| | (11) | 37° 49.1'N 122° 20.7'W | M. edulis | 21°C | Concrete pilings, 1 m above water level |

| Location | Station Initials[a] and Date[b] | Fix | Species | Water Temp. | Description of Site |
|---|---|---|---|---|---|
| San Francisco (9 stations—continuation of NSF 760701) | SSF 760701 (12) | 37° 45.7'N 122° 22.7'W | M. edulis | 19°C | Creosote pier piling |
| | (13) | 37° 43.6'N 122° 04.9'W | M. edulis | 20°C | Wood substrate (apparently not creosote) |
| | (14) | | M. edulis | 22°C | Concrete piling |
| | (15) | | M. edulis | 22°C | Channel marker |
| | (16) | | M. edulis | 23°C | Channel marker |
| | (17) | | M. edulis | 23°C | Creosote wood plyon, main channel |
| | (18) | | M. edulis | 23°C | Channel marker |
| | (19) | | M. edulis | 22°C | Channel marker |
| | (20) | | M. edulis | 22°C | Concrete piling |
| Farallon Islands | FAR 760705 | 37° 41.8'N 122° 59.9'N | M. californianus | 14°C | Horizontal tidal ledge, natural rock, upper range |
| Point Arena | PA 760710 | 38° 57.4'N 122° 59.9'W | M. californianus | 13°C | Erosional sedimentary rock, upper range |
| Cape Mendocino | CM 760612 | 40° 24.3'N 124° 23.4'W | M. californianus | 12°C | Small rocky point, on soft conglomerate rock |
| Humboldt | H 760714 | 40° 45.8'N 124° 14.4'W | M. californianus | 14°C | On jetty rock |
| Point St George | PSG 760716 | 51° 46.7'N 124° 15.2'W | M. californianus | 13°C | Natural rock (shale type), horizontal ledge, upper range |
| Gold Beach, Ore. | GB 760720 | 42° 27.0'N 124° 25.6'W | M. californianus | 12°C | Semi-horizontal natural rock |
| Coos Bay, Ore. | CB 760623 | 143° 19.9'N 24° 22.9'W | M. californianus | 12°C | Horizontal tidal shelf, natural rock |
| Yaquina Head | Y 760726 | 44° 41.0'N 124° 04.7'W | M. californianus | 10°C | On natural rock, upper ranges from small rocky point |
| Tillamook Bay (5 stations) | T 760729 (1) | 45° 31.5'N 123° 53.8'W | M. edulis | 14°C | Man-made rock fill-point, near oyster processing plant |
| | (2) | 45° 32.1'N 123° 53.9'W | M. edulis | 14°C | Flood channel, concrete substrate |
| | (3) | 45° 32.9'N 123° 54.4'W | M. edulis | 13°C | On concrete remnants |
| | (4) | 45°33.5'N 123° 55.1'W | M. edulis | 13°C | Natural rock substrate |
| | (5) | 45° 33.9'N 123° 56.2'W | M. edulis | | Entrance channel to bay |
| Columbia River | CR 760802 | 46° 14.0'N 124° 03'W | M. californianus | 12°C | Jetty rock |
| Willapa Bay (2 stations) | WB 760804 (1) | 46° 40.4'N 123° 55.6'W | M. edulis | 14°C | Natural rock substrate, mussels are underneath rock at mud line |
| | (2) | 46° 43.7'N 124° 04.0'W | M. edulis | 14°C | Area largely mud-flats for both stations |
| Grays Harbor | GH 760806 | 46° 55.7'N 124° 10.5'W | M. californianus | 13°C | Area largely mud-flats, mussels from jetty |
| Cape Flattery | CF 760810 | 48° 23.1'N 123° 43.6'W | M. californianus | 11°C | Horizontal tidal ledge, natural rock substrate |
| Puget Sound | PS 760812 | 47° 54.2'N 122° 22.9'W | M. edulis | 16°C | Natural rock, south end of Whidbey Island |
| Boundary Bay | BB 760818 | 48° 56.5'N 123° 49.2'W | M. edulis | 13°C | Natural rock |
| Blue Hills Falls, Maine | BHF 760915 | 44° 22.5'N 68° 33.5'W | M. edulis | 16°C | Horizontal tide-pool, upper range |
| Sears Island, Maine | SI 760916 | 44° 25.8'N 68° 53.1'W | M. edulis | 17°C | Horizontal tide-flat, small rocky area |
| Cape Newagen, Maine | CN 760917 | 43° 47.2'N 69° 39.2'W | M. edulis | 16°C | Natural rock |

| Location | Station Initials[a] and Date[b] | Fix | Species | Water Temp. | Description of Site |
|---|---|---|---|---|---|
| Bailey Island, Maine | BI 760918 | 43° 44.9′N 69° 59.4′W | M. edulis | 17°C | Rock ledges |
| Portland, Maine | P 760920 | 43° 41.2′N 70° 13.9′W | M. edulis | 20°C | Stone-boulder beach substrate |
| Cape Ann, Mass. | CA 760921 | 42° 39.6′N 70° 37.3′W | M. edulis | 16°C | Tide-pool rocks, adjacent to sandy beach |
| Boston | B 760922 | 42° 21.3′N 71° 58.1′W | M. edulis | 15°C | Small stone beach, northwest corner Deer Island |
| Plymouth, Mass. | P 760923 | 41° 56.5′N 70° 36.7′W | M. edulis | 15°C | Rocky-stone beach |
| Cape Cod Canal, Mass. | CCC 760925 | 41° 44.45′N 70° 36.8′W | M. edulis | 15°C | Granite rock along canal |
| Sakonnet Point, Rhode Island | SK 760927 | 41° 27.2′N 71° 11.7′W | M. edulis | 17°C | Natural rock shelves, between sandy beaches |
| Block Island, Rhode Island | LK 761001 | 41° 11.6′N 71° 35.7′W | M. edulis | 16°C | Small stone beach |
| Millstone, Conn. | MIL 761004 | 41° 18.3′N 72° 09.8′W | M. edulis | 19°C | Natural rock tidal ledges, S. of nuclear power-plant outfall (temp. 28°C near outfall) |
| New Haven, Conn. | NH 761006 | 41° 11.5′N 73° 03.2′W | M. edulis | 17°C | Rock substrate, Charles Island |
| Manhasset Neck, New York | MN 861009 | 40° 42′N 73° 43.9′W | M. edulis | 13°C | Shell-bed substrate on coarse sand to small rock beach |
| Herod Point, Long Island, N.Y. | HP 761010 | 40° 58.1′N 72° 50.7′W | M. edulis | 13°C | Stone-sand substrate beach |
| Fire Island, Long Island, N.Y. | FI 761011 | 40° 46′N 72° 45.1′W | M. edulis | 15°C | Jetty rock |
| Great Gull Island | GGI 761011 | 41° 12.2′N 72° 07.1′W | M. edulis | | |
| Rockaway Point, New York | RK 761012 | 40° 32.5′N 73° 56.5′W | M. edulis | 15°C | Jetty rock substrate |
| Atlantic City, New Jersey | AC 761014 | 39° 22.3′N 74° 24.3′W | M. edulis | | Jetty rock substrate |
| Cape May, New Jersey | MAY 761015 | 38° 55.6′N 74° 57.8′W | M. edulis | | Jetty rock substrate |
| Cape Henlopen, Delaware | CH 761004a | 38° 46.8′N 75° 07.7′W | M. edulis | 11°C | Jetty rock substrate |
| | CH 761104b | 38° 46.8′N 75° 07.7′W | Oysters | | |
| Assateague, Maryland | A 761106 | 38° 19.4′N 75° 05.6′W | M. edulis | 11°C | Jetty rock |
| Cape Charles, Virginia | CC 761109 | 37° 17.3′N 76° 01′W | C. virginica | 5°C | Tidal marshland, substrate is old oyster shells and mud |
| Lynnhaven Bay, Virginia | LH 761115 | 36° 54.2′N 76° 05.3′W | C. virginica | 8°C | Marsh-grass habitat |
| Wachapreague Inlet, Va. | W 76116a | 37° 35.2′N 75° 36.8′W | M. edulis | 6°C | Wood pier pilings |
| | W 76116b | 37° 35.2′N 75° 37.8′W | C. virginica | 6°C | |
| Hatteras Island, North Carolina | H 761119 | 35° 12.3′N 75° 43′W | C. virginica | 12°C | Tidal marshland, mud substrate |
| Beauford, North Carolina | BU 761121 | 34° 43′N 76° 40.7′W | C. virginica | 11°C | Tidal basin |
| Cape Fear, North Carolina | FR 761122 | 33° 57′N 77° 55.5′W | C. virginica | 10°C | Tidal marshland |
| Sapelo Island, Georgia | SAP 761129 | 31° 23.6′N 81° 16.6′W | C. virginica | 12°C | Tidal creek |
| Savannah River, Georgia | SAV 761130 | 32° 01′N 80° 52.4′W | C. virginica | 12°C | Banks of South Channel |

| Location | Station Initials[a] and Date[b] | Fix | Species | Water Temp. | Description of Site |
|---|---|---|---|---|---|
| Charleston, South Carolina | CH 761201 | 32° 44.2'N 79° 52.5'N | C. virginica | 11°C | Fort Johnson |
| St. Augustine, Florida | St.A. 761204 | 29° 42.5'N 81° 13.8'W | C. virginica | 15°C | Tidal marsh |
| Smithsonian Harbor Branch Lab. | SMTH 761218 | | O. equestris | | Raised in laboratory tanks |
| Naples, Florida | NAP 761223 | 26° 01.3'N 81° 44.1'W | O. equestris | | Mangrove roots |
| Anclote, Florida | AN 761230 | 28° 11.4'N 83° 47.5'W | O. equestris | 18°C | Oyster beds north of electric power-plant |
| Cedar Key, Florida | KEY 761231 | 29° 03.5'N 83° 02'W | C. virginica | 18°C | Commercial oyster bed |
| Apalachicola, Florida | AP 770104 | 29° 42.9'N 84° 53.4'W | C. virginica | | Commercial oyster bed |
| Panama City, Florida | PAN 770105 | 29° 42.9'N 84° 54.3'W | C. virginica | | West Bay of St. Andres Bay |
| Pensacola, Florida | PEN 770108 | 30° 20.3'N 87° 09.5'W | C. virginica | 12°C | Sabine Point, Santa Rosa Island |
| Dauphin Island, Alabama | DAU 770113 | 30° 18'N 88° 08.5'W | C. virginica | 10°C | Commercial subtidal reef |
| Pass Christian, Mississippi | PAS 770115 | 30° 18'N 89° 15.8'W | C. virginica | 8°C | Subtidal reef |
| Biloxi, Miss. | BIL 770117 | 30° 23.4'N 88° 51.5'W | C. virginica | 7°C | Shallow waters, next to dock |
| Lake Borgne, Louisiana | BOR 770120 | 29° 57.7'N 89° 39.1'W | C. virginica | | Southwest area of lake |
| Lake Campo, Louisiana | CAM 770123 | 29° 39.8'N 89° 38.7'W | C. virginica | | |
| Black Bay, Louisiana | BLA 770124 | 29° 37'N 89° 37'W | C. virginica | | West area of bay |
| Bastian Bay | BAS 770125 | 29° 19.2'N 89° 38.9'W | C. virginica | | Northern area of bay |
| Bayou de West, Louisiana | DEW 770126 | 29° 12'N 91° 03.4'W | C. virginica | | Between Bay Junop and Gulf of Mexico |
| Lake Barre, Louisiana | BAR 770126 | 29° 15.7'N 90° 34.1'W | C. virginica | | West side of Lake |
| Bay de Ilettes, Louisiana | ILE 770127 | 29° 17.1'N 90° 00.1'W | C. virginica | | In Bay de Illettes, just west of Point Peiro; oil drilling in area |
| Barataria Bay, Louisiana | BRT 770128 | 29° 21'N 89° 55.9'W | C. virginica | | Central area of Bay, near Middle Bank Light |
| Calcasieu Lake, Louisiana | CAL 770201 | 30° 9.5'N 93° 20.2'W | C. virginica | | South end of Lake, just east of ship channel |
| Galveston Bay | GAL 770219 | 29° 29.2'N 94° 46.3'W | C. virginica | | |
| East Matagordo Bay | EMB 770315 | 28° 40.0'N 95° 55.1'W | C. virginica | | |
| Fulton | FUL 770316 | 28° 08.5'N 97° 02.9'W | C. virginica | | |
| San Antonio Bay | ANT 770316 | 28° 18.6'N 96° 40.2'W | C. virginica | | |
| Panther Point | PAN 770316 | 28° 12.5'N 96° 42.0'W | C. virginica | | |
| Matagordo Bay | MAT 770316 | 28° 38.0'N 96° 02.0'W | C. virginica | | |
| Espiritu Santo Bay | ESB 770310 | 28° 18.6'N 96° 36.0'W | C. virginica | | |
| Lake Sabine Bay | SAB 770318 | 29° 43.2'N 93° 51.5'W | C. virginica | | |

[a] Initials of the station's name.

[b] Date of collection, starting with year, then month and day. A small letter or seventh digit in some samples refers to the station sequence.

[c] Explanations are given in the right-hand column.

TABLE 2.2 Concentrations of the Unresolved Complete Mixture (UCM) of Alkanes and Cycloalkanes in Mussels and Oysters

| Station | Unresolved Complete Mixture $10^{-6}$ g/g Dry Weight | | |
| --- | --- | --- | --- |
| | Duplicates | | |
| | A | B | Average |
| *Maine* | | | |
| Blue Hill Falls (BHF) | < 5 | < 5 | < 5 |
| Sears Island (SI) | 44 | 49 | 47 |
| Cape Newagen (CN) | 9 | 8 | 8.5 |
| Bailey Island (BI) | 9 | 8 | 8.5 |
| Portland (P) | 111 | 118 | 115 |
| *Massachusetts* | | | |
| Cape Ann (CA) | < 3 | < 3 | < 3 |
| Boston (B) | 292 | 305 | 298 |
| Plymouth (PL) | 21 | 14 | 17 |
| Cape Cod Canal (CCC) | < 3 | < 3 | < 3 |
| *Rhode Island* | | | |
| Sakonnet Point (SK) | < 3 | < 3 | < 3 |
| Block Island (BLK) | 6.4 | < 3 | ∿ 4 |
| *Connecticut* | | | |
| Millstone (MIL) | 40 | 34 | 37 |
| New Haven (NH) | 30 | 13 | 21 |
| *New York* | | | |
| Manhasset Neck (MN) | 30 | 14 | 22 |
| Herod Point (HP) | 6 | 7 | 6.5 |
| Fire Island (FI) | 22 | 17 | 20 |
| Rockaway Point (RK) | 96 | 68 | 82 |
| *New Jersey* | | | |
| Atlantic City (AC) | 4 | 5 | 4.5 |
| Cape May (MAY) | 18 | 21 | 20 |
| *Delaware* | | | |
| Cape Henlopen (CH) | 7 | 9 | 8 |
| *Maryland* | | | |
| Assateague (A) | 10 | 16 | 12 |
| *Virginia* | | | |
| Cape Charles (CC) | < 3 | < 3 | < 3 |
| Lynnhaven Bay (LH) | 10 | 14 | 12 |
| Wachapreague Inlet (W) | < 3 | < 3 | < 3 |

TABLE 2.2 (Continued)

| Station | Unresolved Complete Mixture $10^{-6}$ g/g Dry Weight | | |
| | Duplicates | | |
| | A | B | Average |
|---|---|---|---|
| *North Carolina* | | | |
| Hatteras Island (H) | < 3 | < 3 | < 3 |
| Beaufort (BU) | 27 | 28 | 27.5 |
| Cape Fear (FR) | 41 | 20 | 30 |
| *South Carolina* | | | |
| Charleston (CHR) | 89 | 97 | 93 |
| *Georgia* | | | |
| Sapelo Island (SAP) | < 3 | < 3 | < 3 |
| Savannah River (SAV) | 33 | 34 | 33.5 |
| *Florida* | | | |
| St. Augustine (ST. A) | 6 | 4 | 5 |

SOURCE: J. W. Farrington, unpublished data, U.S. Mussel Watch Program (1978).

## FOSSIL FUEL COMPOUNDS IN BIVALVES RELATIVE TO SURROUNDING HABITAT

The use of bivalves as sentinel organisms to detect the presence of fossil fuel hydrocarbon contamination depends on the ability of the organism to accurately reflect the hydrocarbon composition of its environment.

Laboratory studies on the dynamics of uptake and depuration of hydrocarbons by oysters (Stegeman and Teal 1973), mussels (Fossato and Canzonier 1976, Fossato 1975), and several species of shellfish (Anderson 1975) yielded the following generalized pattern. When placed in seawater containing sublethal concentrations of hydrocarbons, the animals take up the compounds at an initially high rate and accumulate them to a maximum concentration until either the animals' tissues are saturated or an equilibrium has been reached between the concentration in animal tissues and the surrounding water (Figure 2.7A). When placed in clean water the animals depurate to some background level (Figure 2.7B). Both $R_I$ (initial slope of the hydrocarbon concentration in bivalves as a function of time) and $C_{max}$ (maximum hydrocarbon concentration in bivalves for a given seawater concentration) are dependent on the concentration of

TABLE 2.3  Concentrations of Fluoranthene and Pyrene in Mussels and Oysters (ppm-$10^{-6}$ g/g Dry Weight)[a]

| Station | East and Gulf Coast, 1976–1977[b] | |
|---|---|---|
| | Fluoranthene | Pyrene |
| *Maine* | | |
| Blue Hill Falls (BHF) | .005 | .003 |
| Sears Island (SI) | .047 | .051 |
| Cape Newagen (CN) | .084 | .052 |
| Bailey Island (BI) | .012 | .008 |
| Portland (P) | .064 | .056 |
| *Massachusetts* | | |
| Cape Ann (CA) | .012 | .066 |
| Boston (B) | .240 | .329 |
| Plymouth (PL) | .016 | .014 |
| Cape Cod Canal (CCC) | .012 | .012 |
| *Rhode Island* | | |
| Sakonnet Point (SK)[c] | .012; .014 | .004; .006 |
| Narragansett Bay (N) | .026 | .025 |
| Block Island (BLK)[c] | .037; .039 | .023; .022 |
| *Connecticut* | | |
| Millstone (MIL) | .032 | .028 |
| New Haven (NH) | .034 | .048 |
| *New York* | | |
| Manhasset Neck (MN) | .114 | .381 |
| Herod Point (HP) | .034 | .021 |
| Fire Island (FI) | .020 | .012 |
| Great Gull Island (GGI) | .018 | .016 |
| Rockaway Point (RK) | .067 | .107 |
| *New Jersey* | | |
| Atlantic City (AC) | .035 | .031 |
| Cape May (MAY) | .050 | .051 |
| *Delaware* | | |
| Cape Henlopen (CH) | .032 | .030 |
| *Maryland* | | |
| Assateague (A) | .027 | .018 |
| *Virginia* | | |
| Cape Charles (CC) | .047 | .019 |
| Lynnhaven Bay (LH) | .106 | .062 |
| Wachapreague Inlet (W) | .023 | .012 |
| *North Carolina* | | |
| Hatteras Island (H) | .042 | .023 |

TABLE 2.3  (Continued)

| Station | East and Gulf Coast, 1976–1977[b] | |
|---|---|---|
| | Fluoranthene | Pyrene |
| Beaufort (BU) | .169 | .118 |
| Cape Fear (FR) | .009 | .013 |
| *South Carolina* | | |
| Charleston (CHR) | .616 | .258 |
| *Georgia* | | |
| Sapelo Island (SAP) | .005 | .016 |
| Savannah River (SAV) | .192 | .157 |
| *Florida* | | |
| St. Augustine (ST. A) | .062 | .039 |
| Naples (NAP) | .008 | .007 |
| Anclote (AN) | .033 | .033 |
| Cedar Key (KEY) | .020 | .010 |
| Apalachiola (AP) | .010 | .006 |
| Panama City (PAN) | .036 | .017 |
| Pensacola (PEN) | .070 | .040 |
| *Alabama* | | |
| Dauphin Island (DI) | .018 | .011 |
| *Mississippi* | | |
| Pass Christian (PAS) | .050 | .026 |
| Biloxi (BIL)[c] | .042; .094 | .063; .104 |
| *Washington* | | |
| Boundary Bay (BB) | 3.35 | 1.54 |
| Puget Sound (PS) | N.D. | N.D. |
| Cape Flattery (CF) | 0.34 | 0.45 |
| Grays Harbor (GH) | N.D. | N.D. |
| Willapa Bay (WB) | N.D. | N.D. |
| *Oregon* | | |
| Columbia River (CR) | N.D. | N.D. |
| Tillamook Bay (T) | N.D. | N.D. |
| Yaquina Head (Y) | Trace | N.D. |
| Coos Bay (CB) | N.D. | N.D. |
| Gold Beach (GB) | Trace | Trace |
| *California* | | |
| Point St. George (PSG) | N.D. | N.D. |
| Cape Mendocino (CM) | N.D. | N.D. |
| Hurnboldt (H) | N.D. | N.D. |
| Point Arena (PA) | N.D. | N.D. |
| Bodega Head (BH) | N.D. | N.D. |
| Farallon Islands (FAR) | Trace | Trace |

TABLE 2.3 (Continued)

| Station | West and Gulf Coast, 1977–1978[d] | |
|---|---|---|
| | Fluoranthene | Pyrene |
| N. San Francisco (NSF) | 0.39 | 0.33 |
| San Francisco (SSF) | 5.72 | 4.13 |
| Santa Cruz (SC) | N.D. | N.D. |
| Soberanas Point (SOB) | N.D. | N.D. |
| San Simeon (SS) | Trace | Trace |
| Diablo Canyon (D) | N.D. | N.D. |
| Point Arguello (PT.A) | N.D. | N.D. |
| Rincon Cliffs (RC) | N.D. | N.D. |
| Santa Catalina Island (SC) | N.D. | N.D. |
| Point Fermin (PF) | N.D. | N.D. |
| San Pedro Harbor (SPH) | 5.61 | 4.60 |
| Oceanside jetty (OS) | N.D. | N.D. |
| Point La Jolla (LJ) | N.D. | N.D. |
| San Diego Harbor (SDH) | 1.06 | 1.35 |
| *Louisiana* | | |
| Drum Bay (DRM) | 0.17 | 0.08 |
| Pumpkin Bay (PUK) | N.D. | N.D. |
| Black Bay (BLA) | Trace | N.D. |
| Quarantine Bay QRT) | N.D. | N.D. |
| Golden Meadows (GOD) | Trace | N.D. |
| Junop Bay (JUN) | N.D. | N.D. |
| Calcasien Lake (CAL) | Trace | Trace |
| *Texas* | | |
| Lake Sabine Bay (SAB) | N.D. | N.D. |
| Galveston Bay (GAL) | 0.94 | 1.01 |
| East Matagordo Bay (EMB) | N.D. | N.D. |
| Matagordo Bay (MAT) | N.D. | N.D. |
| Lavaca Bay North (LAVN) | 0.05 | 0.05 |
| Lavaca Bay (LAV) | Trace | Trace |
| San Antonio Bay (ANT) | N.D. | N.D. |
| Mesquite Bay (MEQ) | N.D. | N.D. |
| Aransas Bay (ARN) | Trace | N.D. |
| Brownsville (BRV) | Trace | Trace |

NOTE: N.D. = None detected above 0.01 ppm.
Trace = Approximately 0.01–0.05 ppm.
[a]Determined by quantitative GC-MS.
[b]East and Gulf Coast analyses by laboratory of Dr. John W. Farrington, Woods Hole Oceanographic Institution.
[c]Duplicate extraction, isolations, and analyses.
[d]West and Gulf Coast analyses by laboratory of Dr. Patrick L. Parker, University of Texas Marine Science Institute, Port Aransas, Texas.

SOURCE: J. W. Farrington and P. L. Parker, unpublished data, U.S. Mussel Watch Program (1978).

hydrocarbons in exposure to water. Stegeman and Teal (1973) measured $R_I$ for the uptake of No. 2 fuel oil by oysters and showed $R_I$ proportional to water concentration for water concentrations from 0 to 450 micrograms per liter ($\mu$g/l). Greater concentrations caused the oysters to cease filtering. In addition, $C_{max}$ from exposure to a given hydrocarbon concentration was different in populations of oysters with different lipid contents. When results were reported on the basis of uptake per gram of body lipid, the same concentrations were reached in two different populations. Tissue saturation occurs when the animal can no longer take up the pollutants at a greater rate or reach a higher $C_{max}$.

Uptake and depuration appear to be passive processes resulting from absorption of hydrocarbons from contaminated food and biochemical equilibration between seawater and body lipids. Laboratory investigations and field experiments involving transplanting mussels to areas containing different pollution levels have shown uptake and depuration to be at least a two-stage process. The initial rapid process toward equilibration has been described by a simple

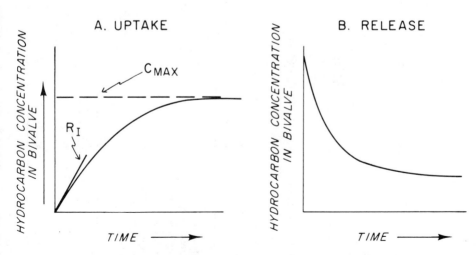

FIGURE 2.7 Stylized representations of the uptake (A) and release (B) of fossil fuel components in bivalves.

TABLE 2.4 Selected Aromatic Hydrocarbon Concentrations in *Mytilus edulis* and *Crassostrea* sp. from East Coast Stations of the U.S. Mussel Watch Program ($10^{-9}$ g/g Dry Weight)

| Compound | Blue Hill Falls, Me. 9/15/76 | Sears Island, Me. 9/16/76 | Portland, Me. 9/20/76 | Boston, Mass. 9/20/76 | Cape Cod Canal, Mass. 9/25/76 |
|---|---|---|---|---|---|
| Naphthalene | 7.6 | 3.2 | 2.6 | 3.9 | 1.8 |
| C-1 Naphthalenes | 9.2 | 2.4 | 8.6 | 29 | – |
| C-2 Naphthalenes | 5.3 | 6.5 | 2.5 | 204 | – |
| Phenanthrenes | 8.1 | 30 | 31 | 91 | 14 |
| C-1 Phenanthrenes | 5.8 | 33 | 34 | 587 | – |
| C-2 Phenanthrenes | – | 61 | 71 | 1820 | 1.6 |
| Fluoranthene | 5.2 | 47 | 64 | 240 | 12 |
| Pyrene | 3.4 | 51 | 56 | 329 | 12 |

| Compound | Sakonnet Point, R.I. 9/27/76 | | Block Island, R.I. 10/01/76 | | Manhasset Neck, N.Y. Rockaway Point |
|---|---|---|---|---|---|
| | A | B[a] | A | B[a] | |
| Naphthalene | – | – | – | – | 2.5 |
| C-1 Naphthalenes | – | – | 8 | 1.6 | 2.5 |

| | | | | | |
|---|---|---|---|---|---|
| C-2 Naphthalenes | – | – | 23 | 17 | – | 7.7 |
| Phenanthrene | 9.9 | 14 | 20 | 18 | 18 | 25 |
| C-1 Phenanthrenes | 6.4 | 5.7 | 33 | 30 | 23 | 32 |
| C-2 Phenanthrenes | – | – | 41 | 44 | 84 | 56 |
| Fluoranthene | 12 | 14 | 37 | 40 | 114 | 67 |
| Pyrene | 4 | 5 | 23 | 22 | 381 | 107 |

| Compound | Sapelo Island, Ga. 11/29/76 | Savannah River, Ga. 11/30/76 | Charleston, S.C. 12/01/76 | Cape Fear, N.C. 11/22/76 |
|---|---|---|---|---|
| Naphthalene | 5 | 4 | 5 | 2 |
| C-1 Naphthalenes | 6 | 6 | 12 | 4 |
| C-2 Naphthalenes | 6 | 28 | 48 | 4 |
| Phenanthrene | 15 | 92 | 85 | 15 |
| C-1 Phenanthrenes | 13 | 86 | 258 | 12 |
| C-2 Phenanthrenes | 5 | 192 | 616 | 8.6 |
| Fluoranthene | 16 | 157 | 258 | 13 |
| Pyrene | 7 | 212 | 251 | 7.6 |

[a] Duplicate analyses of homogenized sample.

SOURCE: J. W. Farrington, unpublished data, U.S. Mussel Watch Program (1978).

exponential function with half-life in body tissues
calculated as approximately 3 to 20 days (Fossato and
Canzonier 1976, Burns and Smith 1978, DiSalvo and others
1975). Time for complete equilibration appears to be on the
order of 90 days under field conditions.

Various physiological mechanisms can be proposed for the
initially fast equilibration of most of the body burden with
surrounding water and the subsequent slower change toward
equilibration. These include concepts of differential
partitioning of hydrocarbons between various tissue
components and the mobility of hydrocarbons incorporated
into various lipid pools.

Mussel tissues appear to saturate at approximately 30 mg
hydrocarbons per gram body lipid. This is the approximate
maximum concentration in animals taken from a variety of
polluted areas (Burns and Smith 1977).

Analysis of seawater and mussels sampled from the same
location show that if the hydrocarbons are present in a
"dissolved form" (those that pass through a glass fiber
filter and onto a resin absorption column), then the pattern
obtained by gas chromatography of mussel extracts will look
qualitatively similar to a chromatogram of the seawater
extract (Risebrough and others 1979). If the hydrocarbons
are present primarily in "particulate" form (those trapped
on a glass filter), then the mussels will show a relative
enrichment of the lower-boiling compounds in both saturated
and unsaturated hydrocarbon fractions (Figure 2.8)
(Risebrough and others 1979).

Most hydrocarbons in seawater (except in moderately
polluted areas) appear to be present in particulate form.
Thus the qualitative pattern of hydrocarbons in mussels is
usually of slightly lower boiling range than that of
seawater from the same area (Burns and Smith 1978,
Risebrough and others 1979).

Bivalves appear to have minimal ability to metabolize
petroleum hydrocarbons (Vandermeulen and Penrose 1978,
Bayne, Institute for Marine Environmental Research,
Plymouth, England, personal communication, December 1978),
and therefore do not extensively modify the pattern of
hydrocarbons incorporated into body tissues. This
observation adds further support to the usefulness of
mussels for reflecting the qualitative composition of the
water column to which they are exposed.

To estimate the amount of time required by mussels to
equilibrate to ambient water concentrations in the field,
Burns and Smith (1978) selected two coastal bay stations in
Australia that had relatively consistent but different

pollution concentrations.  Mussels were collected at each
station, placed in polypropylene mesh bags, and transferred
to the opposite site.  After 51 days both native and
transplanted mussels were collected.  Results of hydrocarbon
analyses on the samples are shown in Table 2.5.  Levels of
oil pollution vary slightly at both sites, but the data
indicate that neither set of transplanted mussels had
reached equilibrium with ambient water levels after 51 days.
Mussels taken from Site I (high pollution) that had been
moved to Site II (moderate pollution) were still depurating.
Mussels from Site II that had been moved to Site I were
still taking up hydrocarbons.  Gas chromatograms of the
samples showed that Site I was contaminated with a series of
substituted benzene derivatives that could be distinguished
from the large background of degraded crude oil.  The
aromatic content of Site I samples was much higher than that
of samples from Site II.  The data show that mussels tend to
equilibrate with hydrocarbon concentrations in ambient water
(Table 2.5).  The average rates of change in hydrocarbon
content were the same for both uptake and depuration
processes.  Time to reach equilibrium was estimated to be
approximately 90 days.  However, since the rate of change in
concentration is slower as equilibrium is approached, the
time is a minimum estimate.

Long-term site monitoring was conducted near a refinery
pier in Westernport Bay (Australia).  Tidal mixing is
extremely rapid in this area and other monitoring results
have shown that the refinery effluent is well mixed in
receiving waters (Smith and Burns 1978a).  Table 2.6 shows
that concentrations of hydrocarbons in mussels from the pier
were consistent when expressed on a lipid weight basis
regardless of season, with a few obvious exceptions.
Unusually low values were obtained on November 2, 1977, when
the refinery had shut down for maintenance.  Other
exceptions were high values resulting from small oil spills
(October 26, 1977 and September 21, 1978).  Preliminary
evidence on selective size-class analysis showed no
consistent trend in body concentration with size (Table
2.7).  Additional data from a variety of areas should be
obtained to clarify the effects of size, age, seasonality,
or other parameters on body concentration.

Several authors have proposed models to explain the
bioconcentration of hydrocarbons in mussel tissues based on
equilibration of hydrocarbons between animal lipids and
surrounding waters.  The simplest model assumes
bioconcentration is independent of other constituents in
seawater and defines a bioconcentration factor ($K_{Bcf}$) at

32

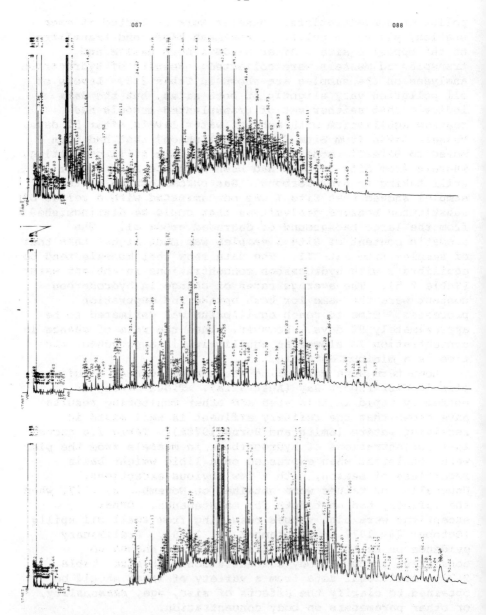

FIGURE 2.8  Hydrocarbons in sea water, sea water particulates, and mussels from San Nicholas Island, California coast, U.S.A. (Reprinted with permission from Proceedings, International Congress on Analytical Techniques in Environmental Chemistry, J. Albaiges, ed., Risebrough and others, Pattern of hydrocarbon contamination in California coastal waters, copyright 1979, Pergamon Press, Ltd.)

Bodega Marine Laboratory Data Partitioning Study, San Nicholas Island, 10-11 September 1978.

equilibrium as the concentration of hydrocarbons in animal tissues divided by concentration in surrounding water.

To test the model, samples of mussels were collected at various stations in Australian coastal waters at the same time water samples were collected. Results are presented in Table 2.8. $K_{Bcf}$ was calculated assuming the mussels were in equilibrium with surrounding water. It appeared from these field data that $K_{Bcf}$ ranges from $10^5$ to $10^7$ based on total hydrocarbons in the $^{14}C$ to $C_{34}$ molecular weight range in mussel lipids (Burns and Smith 1978). Results of similar measurements taken off the California coast were reported by Risebrough and others (1979) and are shown in Table 2.9. The variability in these data may be caused by particular episodes or may result from the fact that $K_{Bcf}$ is dependent on other constituents in seawater. The preliminary data suggested that in areas where seawater contains very low concentrations of hydrocarbons (or perhaps other organics), $K_{Bcf}$ is much higher than areas where pollution is moderate to heavy. Before the analysis of mussels can be used to reflect quantitatively the amount of hydrocarbons in ambient

---

*Top:* Polyurethane foam extract of the "dissolved" phase of seawater
Volume 1,276 liters, 2.0 $\mu$l injection, 23 $\mu$l extract volume.
Att $2^{10}$, $2^4$ @ 9 minutes.
Total resolved hydrocarbons:     4.0 ng/liter
Total unresolved hydrocarbons: 13.0 ng/liter
Total alkanes:                            2.4 ng/liter

*Middle:* Seawater particulates
Volume 2,863 liters, 2.0 $\mu$l injection, 90 $\mu$l extract volume.
Att $2^{10}$, $2^4$ @ 9 minutes.
Total resolved hydrocarbons:     8.5 ng/liter
Total unresolved hydrocarbons: 12.0 ng/liter
Total alkanes:                            3.0 ng/liter

*Bottom: Mytilus californianus*
102 grams wet weight, 2.0 $\mu$l injection, 82 $\mu$l extract volume.
Att $2^{10}$, $2^4$ @ 9 minutes.
Total resolved hydrocarbons:     3.0 $\mu$g/gram wet weight
Total unresolved hydrocarbons: 4.0 $\mu$g/gram wet weight
Total alkanes:                            0.31 $\mu$g/gram wet weight

All samples were analyzed on a Hewlett-Packard 5840A gas chromatograph equipped with a 7176 automatic liquid sampler, 18835 capillary inlet system, 30 meter (0.25 mm ID) SP-2100 glass capillary column (J & W Scientific) and flame ionization detector.

Initial temperature 65°C, 2 minutes
Rate 3.5°/minute
Final temperature 270°C

TABLE 2.5 Hydrocarbon Content of Mussels from Two Monitoring Sites in Australia and Mussels Transplanted between the Two Sites

| | Hydrocarbon Content | | | | |
|---|---|---|---|---|---|
| | Saturates | Unsaturates | Total | | 6-Substituted Benzene Derivatives (mg/g lipid)[a] |
| Sample & Date | mg/g lipid | mg/g lipid | μg/g dry | mg/g lipid | |
| Site II 12/19/77 | 4.6 | 2.3 | 837 | 6.9 | – |
| Trans from Corio | 2.8 | 2.8 | 818 | 5.6 | 0.3 |
| Site II 2/10/78 | 2.0 | 1.3 | 454 | 3.3 | – |
| Site I 12/19/77 | 3.4 | 6.0 | 1,421 | 9.4 | 1.2 |
| Trans from Site II | 4.2 | 5.3 | 1,354 | 7.5 | 0.9 |
| Site I 2/10/78 | 4.2 | 6.4 | 1,700 | 10.6 | 1.3 |

[a]Special Contaminants in Corio samples making up part of the unsaturated hydrocarbons.

SOURCE: Burns and Smith (1978).

TABLE 2.6 Petroleum Hydrocarbon Content of Mussels from Refinery Wharf in Westernport Bay (Australia) Expressed on Both Dry-Weight and Lipid-Weight Basis

| Date | Lipid Wt. % of Dry Wt. | Total Petroleum Hydrocarbons | | Unsaturated Hydrocarbons | |
|---|---|---|---|---|---|
| | | µg/g dry | mg/g lipid | µg/g dry | mg/g lipid |
| 7/4/75 | 16.8 | 735 | 4.4 | 220 | 1.3 |
| 12/12/75 | 15.0 | 689 | 4.5 | 206 | 1.3 |
| 2/12/76 | 15.0 | 570 | 3.8 | 171 | 1.1 |
| 4/28/76 | 11.9 | 685 | 3.5 | 205 | 1.0 |
| 4/28/76 (subtidal) | 15.7 | 1,186 | 6.4 | 364 | 2.0 |
| 11/2/77 | 10.9 | 158 | 1.4 | 81 | 0.7 |
| 11/2/77 (subtidal) | 8.2 | 156 | 2.1 | 86 | 1.2 |
| 4/29/77 | 15.5 | 660 | 4.2 | 198 | 1.3 |
| 6/8/77 | 17.6 | 970 | 5.4 | 308 | 1.7 |
| 10/26/77 | 15.9 | 1,975 | 12.3 | 668 | 4.2 |
| 12/19/77 | 11.0 | 837 | 6.9 | 280 | 2.3 |
| 2/10/78 | 14.0 | 454 | 3.3 | 182 | 1.3 |
| 9/21/78 | | 1,241 | 8.2 | 541 | 3.6 |
| 9/25/78 | | 721 | 5.5 | 352 | 2.7 |
| 10/3/78 | | 603 | 4.9 | 280 | 2.3 |
| 10/11/78 | | 529 | 4.5 | 177 | 1.5 |

SOURCE: Burns and Smith (1978).

TABLE 2.7  Total Hydrocarbon Content in Different Size Classes of Mussels Sampled Simultaneously from the Same Station

| Site | $\mu$g/g dry | mg/g lipid |
|------|--------|-----------|
| *Site A  3/25/76* | | |
| (avg. 7.8 cm length) | 989 | 6.6 |
| (avg. 5.1 cm length) | 837 | 5.6 |
| | | |
| *Site B  10/6/77* | | |
| (avg. 6.0 cm length) | 1,423 | 12.4 |
| (avg. 3.4 cm length) | 1,703 | 14.9 |

SOURCE: K. A. Burns, panel member, unpublished data, Australia (1978).

water, further research must be undertaken to determine what factors govern $K_{Bcf}$.

Analyses of mussels or other bivalves can provide valuable information on relative distributions of hydrocarbons in the water column. Since the animals reflect the hydrocarbon concentrations in the waters to which they are exposed, they can be used to estimate sources and types of contaminants. With caution, such analyses can also be used to estimate relative water-column concentrations of pollutants. Long-term monitoring should indicate a base pattern of contamination, provided water concentrations are relatively uniform. In areas of fluctuating inputs, analyses will show greater variation, with sudden inputs showing up as spikes over the average concentration.

Since complete equilibration takes several months and since variation in results seems to be reduced by expressing results on a lipid basis, sampling of indigenous populations of bivalves appears feasible. Because the bioconcentration potential of hydrocarbons in mussel tissues is high, mussel analyses are most useful in assessing water quality in the low ppb range for total hydrocarbons. The existence of areas polluted by visible slicks and oil films would be apparent in mussels, but tissue saturation would limit the usefulness of mussels to reflect relative water quality quantitatively.

Mussels can be used to indicate areas of elevated environmental concentrations of hydrocarbons. Once problem areas are identified, further assessment of environmental quality can be achieved by analysis of sediments and seawater, together with biochemical and biological assessments of organism and community health.

TABLE 2.8   Hydrocarbon Content of Mussels and Sea Water Sampled
Simultaneously in Three Areas of Victorian Coastal Waters

| Sample Station and Date | Hydrocarbons | | |
|---|---|---|---|
| | Water ($\mu$g/1) | Mussels (mg/g lipid) | $K_{Bcf}$[a] |
| *CORIO BAY* | | | |
| Shell Beacon 9 9/28/77 | 22.6 | 12.1 | $5.3 \times 10^5$ |
| Silo Beacon 4 9/28/77 | 1.0 | 3.3 | $3.3 \times 10^6$ |
| Ripp. Beacon 1 9/27/77 | 0.2 | 2.7 | $13.5 \times 10^6$ |
| Hope. Beacon 15 9/27/77 | 1.0 | 3.1 | $3.1 \times 10^6$ |
| AVERAGE | | | $5.1 \times 10^6$ |
| *HOBSONS BAY* | | | |
| Webb. Beacon 18 10/6/77 | 9.7 | 12.5 | $1.1 \times 10^6$ |
| Pt. Melb. Bcn. 8 10/4/77 | 8.6 | 4.8 | $5.5 \times 10^5$ |
| P. Pile Altona 10/11/77 | 0.3 | 0.6 | $2.0 \times 10^6$ |
| Fawkner Beacon 10/4/77 | 1.9 | 0.3 | $1.6 \times 10^5$ |
| AVERAGE | | | $1 \times 10^6$ |
| *WESTERNPORT BAY* | | | |
| Stony Pt. 10/24/77 | 0.3 | 2.7 | $9.0 \times 10^6$ |
| Esso Wharf 10/27/77 | 0.7 | 3.9 | $5.5 \times 10^6$ |
| B.P. Wharf | | | |
| 10/26/77 | 0.8 | 12.4 | $1.5 \times 10^7$ |
| 9/21/78 | 0.5 | 8.2 | $1.6 \times 10^7$ |
| 9/25/78 | 0.25 | 5.5 | $2.5 \times 10^7$ |
| 10/3/78 | 0.15 | 4.9 | $4.9 \times 10^7$ |
| 10/11/78 | 0.15 | 4.5 | $4.5 \times 10^7$ |
| AVERAGE | | | $2.3 \times 10^7$ |

[a] $K_{Bcf}$ calculated assuming equilibrium. Overall average $K_{Bcf} = 4.84 \times 10^6$.

SOURCE: Burns and Smith (1978).

TABLE 2.9 Partitioning of Hydrocarbons among the Mussel (*Mytilus californianus*) and the "Dissolved" and Particulate Phases of Sea Water[a]

| Phase | Palos Verdes | Goleta Point | San Nicholas Island |
|---|---|---|---|
| **DDE** | | | |
| "dissolved" phase | $0.18 \pm .02$ ($6 \times 10^6$) | $0.073 \pm .024$ ($3 \times 10^5$) | $0.044 \pm .007$ ($4 \times 10^5$) |
| particulates | $.082$ ($1 \times 10^7$) | $.029$ ($8 \times 10^5$) | $.008$ ($2 \times 10^6$) |
| mussels | 1,100 | 22 | 18 |
| **PCB** | | | |
| "dissolved" phase | $0.13 \pm .03$ ($3 \times 10^6$) | $0.074 \pm .011$ ($4 \times 10^5$) | $0.10 \pm .04$ ($6 \times 10^4$) |
| particulates | $.044$ ($1 \times 10^7$) | $.019$ ($1 \times 10^6$) | $.011$ ($5 \times 10^5$) |
| mussels | 440 | 27 | 6 |
| **Pristane** | | | |
| "dissolved" phase | $0.061 \pm .03$ ($7 \times 10^5$) | $0.036 \pm .011$ ($2 \times 10^7$) | $0.20 \pm .00$ ($6 \times 10^5$) |
| particulates | $2.0$ ($2 \times 10^4$) | $5.2$ ($1 \times 10^5$) | $2.2$ ($6 \times 10^4$) |
| mussels | 40 | 640 | 120 |
| **Squalene** | | | |
| "dissolved" phase | $0.74 \pm .35$ ($7 \times 10^6$) | $0.25 \pm .07$ ($7 \times 10^6$) | $0.35 \pm .16$ ($4 \times 10^6$) |
| particulates | $32$ ($2 \times 10^5$) | $4.9$ ($4 \times 10^5$) | $30$ ($4 \times 10^4$) |
| mussels | 4,900 | 1,700 | 1,300 |
| **Total Unresolved Saturates** | | | |
| "dissolved" phase | $4.1 \pm 1.3$ ($3 \times 10^7$) | $5.6 \pm 1.2$ ($7 \times 10^7$) | $12 \pm 2$ ($3 \times 10^5$) |
| particulates | $83$ ($2 \times 10^6$) | $970$ ($4 \times 10^5$) | $13$ ($3 \times 10^5$) |
| mussels | 140,000 | 390,000 | 3,900 |

[a]Concentrations in "dissolved" (N=2, with standard deviation) and particulate phases in ng/liter, concentrations in mussels in ng/g. Southern California Bight, August-September, 1978. Partition coefficients ratio of ng/kg in mussels to ng/liter in seawater) are given in parentheses.

SOURCE: Reprinted with permission from Proceedings, International Congress on Analytical Techniques in Environmental Chemistry, J. Albaiges, ed., Risebrough and others, Pattern of hydrocarbon contamination in California coastal waters, copyright, 1979, Pergamon Press, Ltd.

Two monitoring studies have used mussels collected from
"clean" areas, placed in cages, and transferred to areas of
monitoring interest.  Using this approach, Smith and Burns
(1978a) investigated a refinery effluent discharging to
Westernport Bay (Australia).  Phelps and Galloway (1979)
monitored the decrease in hydrocarbon contamination in
Narragansett Bay as stations were sampled with increasing
distance from the Providence River (United States).  Thus,
bivalve transplantation and subsequent analysis can be used
to investigate distributions of contamination from known
sources once problem areas have been identified by initial
screening.

A more detailed knowledge of the quantitative
relationship between concentrations of fossil fuel compounds
in body tissues and surrounding waters constitutes one of
the priority research needs.

## CONCENTRATION OF PETROLEUM HYDROCARBONS IN MUSSELS
RELATIVE TO CONCENTRATION IN SEDIMENTS

Some data are available to allow an assessment of the
usefulness of shellfish as a monitor for the accumulation of
petroleum hydrocarbons in marine sediments and to describe
in general the relationship between mussel concentrations
and sediment concentrations.  This section will outline
several specific cases and will describe whether
measurements of petroleum hydrocarbons in mussels have any
diagnostic value in assessing contamination of sediments and
the biological availability of the contaminants.  As a
preliminary indication of usefulness, it is necessary to
show some consistent relationships between amounts of
petroleum hydrocarbons in bivalves and in sediment.
Elevated concentrations of hydrocarbons in sediment should
be reflected by elevated concentrations in shellfish.

The relationship may be described by an assessment of
fluorescent material in pyrene units in sediment and
shellfish (Eaton and Zitko 1979).  Concentrations found in
harbor areas close to chronic creosote input from wharf
pilings were compared with concentrations in areas remote
from the direct influence of creosote.  Concentrations in
sediment were found to be approximately two orders of
magnitude greater than concentrations in bivalves.  There
does appear to be an increase in concentration in bivalves
when large differences in sediment concentrations (one or
two orders of magnitude) are encountered.  However, when the
individual data are examined, a great deal of variability

exists, and changes smaller than an order of magnitude in sediment hydrocarbon levels are not consistently reflected in bivalve concentrations (Table 2.10).

In the same study, Eaton and Zitko applied the more accurate quantitative analytical method of gas chromatography/mass spectrometry (GC/MS) to quantify six specific hydrocarbons: acenaphthene, fluorene, phenanthrene, fluoranthene, pyrene, and benz(a)anthracene. Only the latter four compounds were found consistently in samples, and although the method was much more accurate, no consistent relationship between sediment concentrations and concentrations in bivalves was shown, except for the fact that "control" areas remote from chronic creosote input have significantly lower levels of the hydrocarbons in sediments and mussels than do wharf areas (Table 2.11). Here differences are at least an order of magnitude, and again it is apparent that bivalves (mussels) would be useful in indicating the gross differences in sediment petroleum hydrocarbon concentration but not in identifying fluctuations of less than one or two orders of magnitude. Table 2.12 contains a sample of selected data showing the lack of relationship between hydrocarbon concentration in sediments and mussels when minor fluctuations in sediment hydrocarbons are involved. The lack of consistency in relative concentrations of petroleum hydrocarbons in sediment and associated bivalves is indicated in data from Gilfillan and Vandermeulen (1978). Five-fold increases in sediment concentrations are not reflected by any relative increases in concentrations in clams.

TABLE 2.10   Concentration of Pyrene in Samples from Eastern Canada (pyrene units $/\mu g/ml \times 10^6$)

| Station | Sediment | Clams | Mussels |
| --- | --- | --- | --- |
| A1 | 79,375 | 230 | 115 |
| A2 | 6,645 | 130 | 255 |
| A3 | 37,500 | 58 | 270 |
| B1 | 95,650 | 140 | 175 |
| B2 | 32,975 | 54 | – |
| B3 | 30,990 | 435 | 160 |
| C1 | 14,575 | 37 | 61 |
| C2 | 128,600 | 27 | 61 |
| C3 | 37,825 | 33 | 25 |

SOURCE: Eaton and Zitko (1979).

TABLE 2.11  Aromatic Hydrocarbon Locations (concentration in $\mu$g/g lipid, $\mu$g/g wet weight, geometric mean)

Near Creosoted Wharf Structures

| | n | Phenanthrene | Fluoranthene | Pyrene | Benz(a)anthracene[a] |
|---|---|---|---|---|---|
| Sediment | 26 | 4000 (1.4) | 6190 (2.3) | 4880 (1.8) | 6040 (1.9) |
| Mussels | 17 | 12.7 (0.3) | 48.4 (0.8) | 32.1 (0.5) | 44.4 (0.7) |
| Clams | 2 | 92.8 (0.2) | 85.4 (0.3) | 63.3 (0.3) | 65.0 (0.2) |
| Periwinkles | 39 | 205 (3.2)* | 641 (10.0) | 504 (7.9) | 407 (6.5) |

*n = 35.

Remote from Creosoted Wharf Structures

| | n | Phenanthrene | Fluoranthene | Pyrene | Benz(a)anthracene[a] |
|---|---|---|---|---|---|
| Sediment | 5 | 22.2 (.010) | 17.1 (.008) | 10.4 (.007) | 11.7 (.005) |
| Mussels | 2 | 0.5 (.006) | 0.9 (.005) | 0.4 (.002) | ND (.020)* |
| Clams | 2 | 1.0 (.003) | 1.0 (.006) | 0.8 (.003) | 0.9 (ND) |
| Periwinkles | 2 | 0.7 (.003) | 0.5 (.003) | 0.2 (.004) | ND (.002) |

*n = 1.

[a]Compounds of formula $C_{18}H_{12}$ including chrysene, benzanthacenes, and triphenylene were quantified as benz(a)anthracene.

SOURCE: Eaton and Zitko (1979).

TABLE 2.12 Comparison of Aromatic Hydrocarbons in Mussels and Sediments ($\mu$g/g of lipid)

| | Phenanthrene | Fluoranthene | Pyrene | Benzo(a)anthracene |
|---|---|---|---|---|
| Sediment (A)[a] | 1314 | 1304 | 731 | 1273 |
| Mussel (A) | 8.1 | 36 | 30 | 96 |
| Sediment (B) | 586 | 881 | 837 | 1538 |
| Mussel (B) | 36.5 | 46 | 43.6 | 56 |
| Sediment (C) | 5039 | 8334 | 8777 | 7533 |
| Mussel (C) | 95.5 | 123 | 89.3 | 97.5 |
| Sediment (D) | 110 | 65 | 59 | 51 |
| Mussel (D) | 0.7 | 0.9 | 0.6 | 0.6 |
| Sediment (E) | 15414 | 18937 | 15283 | 20960 |
| Mussel (E) | 8.1 | 146 | 91 | 162 |
| Sediment (F) | 3752 | 3850 | 3167 | 3850 |
| Mussel (F) | 14 | 43 | 28 | 49 |

[a] Letters indicate station designations.

SOURCE: Eaton and Zitko (1979).

It is evident from the work described above and from uptake studies that mussels concentrate petroleum hydrocarbons at a much lower concentration than do sediments. Sediments can act as long-term sinks, whereas shellfish reach a saturation point determined by their lipid content. This fact, plus the tremendous variability of sediment levels due to differences in sediment type, hydrographic dynamics, and conditions, make it difficult to determine a useful relationship between hydrocarbon concentrations in sediments and in mussels, particularly where low concentrations are encountered.

In another situation in Westernport Bay, Australia (Burns and Smith 1978), petroleum hydrocarbons were found in mussels while none were detected in adjacent sediment. This situation is probably due to chronic low-level input from a nearby refinery that is rapidly dispersed without settling into the sediments. The opposite situation exists in Chedabucto Bay, Nova Scotia where petroleum hydrocarbons from the "Arrow" spill still contaminate sediments but are not detected in Mytilus edulis (Vandermeulen and Gordon 1976).

If petroleum hydrocarbon concentrations in sediments are of particular importance because of concerns over uptake by food organisms or because of the problems that might arise from the need to dredge the sediment, then it would be more appropriate to monitor a burrowing bivalve such as Mya

<u>arenaria</u>.  For example, in Chedabucto Bay, although no
petroleum hydrocarbons were measured in <u>Mytilus</u>, levels as
high as 95 to 269 ppm were found in interstitial water, and
rooted marine plants (<u>Zostera</u>) had levels up to 25 parts per
trillion (ppt) wet weight.  In this case the clam, <u>Mya
arenaria</u>, was found to have levels as high as 20 ppt.

For monitoring petroleum hydrocarbons in sediments, use
of bivalves is satisfactory only in cases of gross
contamination.  In situations where the environment is being
heavily polluted by petroleum hydrocarbons, elevated levels
in bivalves can be expected.  However, in areas of low
chronic input or in situations where oil spills have
occurred in the past and hydrocarbons have become
incorporated into a fairly stable sediment, the petroleum
hydrocarbon content of bivalves is unlikely to reflect
concentrations in the sediment.


### INTERCALIBRATION NEEDS

Many laboratories, using a variety of analytical
techniques of differing levels of sophistication, are
currently involved in the measurement of fossil fuel
hydrocarbons in marine biota.  The large number of marine-
biota analyses performed in a monitoring program such as the
Mussel Watch Program necessitates the existence of a common
basis for comparing the data.  At present, there is little
or no knowledge of the comparability of hydrocarbon data
from different laboratories.

In setting up interlaboratory comparison exercises,
three types of materials of increasing scientific value and
increasing sophistication are possible: (1) sample splits on
field samples, (2) interim calibration materials, and (3)
reference materials.

<u>Sample splits</u> are of limited value, since they usually
involve collaboration between a few (often only two)
laboratories.  <u>Interim calibration materials</u> are field
samples that have been collected in sufficient quantity to
permit wide distribution and that have been homogenized as
well as possible.

The most sophisticated, and the most reliable,
intercalibration material is a <u>reference material</u> that has
been certified to contain a known amount of certain
constituents.  The National Bureau of Standards (NBS,
Washington, D.C.) provides numerous standard reference
materials (SRMs) that are useful for the calibration of

analytical methods for measuring trace elements in various natural materials.

The formulation of an SRM requires the use of one of three modes of measurement to assure the accuracy of the measurements of concentrations. These modes are: (1) definitive methods, (2) reference methods, and (3) two or more independent and reliable methods. The current state of the art of trace-organic analysis does not permit the issuance of an SRM for trace-level organics in a natural matrix like mussel tissue or sediment. Definitive methods do not exist. Before SRMs can be issued, problems associated with sample homogeneity, storage stability, and matrix effects must be resolved with environmental hydrocarbon determinations.

At present, the analyst ascertains the accuracy of his measurements through the use of internal standards and participation in interlaboratory comparisons on interim calibration samples. The use of internal standards is currently the most reliable means of obtaining accuracy in such measurements.

Recently an interlaboratory comparison for hydrocarbons in mussels was conducted (Wise and others 1979). Two mussel samples, one from a pristine environment in Alaska and one from Santa Barbara, California near natural oil seeps, were homogenized and then analyzed for hydrocarbons by eight laboratories. The laboratories were asked to report the following data: total extractable hydrocarbons, total hydrocarbons in the GC elution range, pristane/phytane ratio, percent water, identities and concentrations of the three most abundant aliphatic and aromatic hydrocarbons, most abundant PAHs greater than four rings, and any additional identification and concentrations made on the sample. Homogeneity studies on the mussel homogenate found relative standard deviations of 14 and 18 percent. The results for three of the requested parameters are summarized in Table 2.13 for the Alaska sample. The ranges of values obtained in different laboratories are generally within factors of 2 to 4. Measurements of individual compounds were in good agreement for most of the laboratories (see Table 2.14). However, in both samples the ranges for some individual n-alkanes differed by as much as an order of magnitude. The differences between laboratories cannot be attributed to sample inhomogeneity.

Recent results by Dunn (1976) on the reproducibility of determinations of BaP indicate the precision attainable for the analyses of mussel homogenates. Tissue from 200 to 300 mussels from several sites was homogenized and then

TABLE 2.13  Hydrocarbon Analyses Content in Alaskan Mussels ($\mu$g/g)

| Laboratory | Total Hydrocarbons in GC Elution Range | | | | | | | Total Extractable Hydrocarbons ($\mu$g/g) | Pristane / Phytane |
|---|---|---|---|---|---|---|---|---|---|
| | Aliphatic | | | Unsaturated/Aromatic | | | Total | | |
| | Resolved | UCM[a] | Total | Resolved | UCM | Total | | | |
| 1 | 4.4 | 14.8 | 19.2 | 8.8 | 29.9 | 38.7 | 57.8 | 139.1 | 58 |
| 2 | | | –[b] | | | | – | – | 40 |
| 3 | | | – | | | | 41.1 | 73.3 | 40.5 |
| | | | – | | | | 32.2 | 38.4 | 38.4 |
| 4 | – | | 12.0 | – | | 89.4 | 101.4 | – | 70.8 |
| 5 | 4.9 | | | | | | – | – | 40.1 |
| | 5.2 | | | | | | – | – | 39.7 |
| 6 | 1.5 | | | 2.1 | – | | – | – | – |
| | 2.6 | | | 2.3 | – | | – | – | – |
| 7 | 3.4 | 16.0 | 19.4 | | | | – | – | 45 |
| | 1.5 | 22.0 | 23.5 | | | | – | – | – |
| 8 | | | | | | | 43.2 ± 18.9 | | |
| Range | 1.5-4.9 | 14.8-22.0 | 12.0-23.5 | 2.1-88 | | 38.7-89.4 | 32.2-101.4 | 38.4-139.1 | 38.4-70.8 |

[a] Unresolved complex mixture in gas chromatogram (UMC).
[b] Dashes indicate that data were not reported for these categories.

SOURCE: Wise et al. (1979).

TABLE 2.14 Aliphatic Hydrocarbon Content in Alaskan Mussels ($\mu g/g$)

| Laboratory | $n\text{-}C_{14}$ | $n\text{-}C_{15}$ | $n\text{-}C_{16}$ | $n\text{-}C_{17}$ | Pristane | Phytane | $n\text{-}C_{22}$ | $n\text{-}C_{23}$ | $n\text{-}C_{24}$ | $n\text{-}C_{25}$ |
|---|---|---|---|---|---|---|---|---|---|---|
| 1 | 0.1 | 0.2 | 0.2 | –[a] | 2.9 | 0.05 | 0.003 | 0.02 | 0.01 | 0.03 |
| 2 | 0.24±0.02 | 0.36±0.03 | 0.37±0.15 | 0.24±0.01 | 2.51±0.34 | ~0.05 | – | – | – | – |
| 3 | – | 0.18 | 0.23 | 0.06 | 3.24 | 0.08 | 0.18 | 0.09 | – | 0.23 |
| 4 | 0.18 | – | 0.45 | – | 1.98 | 0.03 | 0.56 | – | 0.35 | 0.22 |
| 5 | 0.25 | 0.37 | 0.27 | 0.26 | 2.75 | 0.07 | <0.002 | 0.04 | <0.002 | 0.026 |
| 6 | – | – | – | – | – | – | 0.09 | 0.08 | 0.13 | 0.22 |
| 7 | – | 0.25 | 0.20 | 0.21 | 2.34 | 0.052 | – | 0.037 | 0.030 | 0.051 |

[a] Dashes indicate that data were not reported for these categories.

SOURCE: Wise et al. (1979).

subsamples of 20 to 30 g of tissue were analyzed. The results of this study are summarized in Table 2.15. The standard deviation of the analytical results ranged from 3.2 to 8.1 percent of the mean, with an average of 6.2 percent. Samples stored at −10°C for 12 weeks showed no statistically significant change in the content of BaP, a fact that suggests storage stability for at least several months.

The results of intercomparison of mussel analyses were encouraging. However, they should also serve as a warning against overinterpretation of hydrocarbon measurements. Differences of a factor of 4 in data from various laboratories may not be environmentally significant because they may reflect only systematic differences in analyses from different laboratories.

Another observation from the intercalibration exercises is the lack of capability of many laboratories to measure trace amounts of PAHs, many of which may be the most toxic of the fossil fuel hydrocarbons. There exists a need for the development and further application of analytical capabilities in this area.

An important and necessary part of any monitoring program for hydrocarbons in tissue is continuous participation of laboratories in intercomparison exercises to assess the reliability of their data. To improve such an intercalibration program over previous studies, the following recommendations are proposed.

TABLE 2.15 Precision of Assay for Benzo(a)pyrene

| Area | B(a)P level, μg/kg wet weight | | | |
| | Mean | SD | n[a] | SD as % |
| --- | --- | --- | --- | --- |
| Styrofoam floats, Marina A | 19.9 | 1.35 | 6 | 6.8 |
| Untreated wood floats, Marina B | 11.9 | 0.38 | 4 | 3.2 |
| Concrete, 2–4 m from group of 30 creosoted pilings | 18.4 | 1.24 | 6 | 6.7 |
| As above, stored 12 weeks at −10° | 17.8 | 1.45 | 4 | 8.1 |

[a]Subsamples of a homogenate of tissue from 200 to 300 mussels from each location.

SOURCE: Reprinted with permission from Environ. Sci. Technol., vol. 10, B. P. Dunn, Techniques for determination of benzo(a)pyrene in marine organisms and sediments, copyright 1976, American Chemical Society.

- One laboratory should be designated as responsible for initiating and conducting such exercises. This procedure will reduce the work burden created by numerous small interlaboratory studies initiated by several different laboratories.
- A large sample should be prepared in order to conduct a long-term intercalibration program, providing samples for analyses beyond a single exercise.
- In order to provide useful feedback to the participation laboratories, workshops should be held regularly to discuss methodologies and analytical differences in order to investigate their sources.
- Some method of encouraging participation of the various laboratories in these exercises should be employed, such as funding to support the cost of these analyses or a mandatory participation requirement of all laboratories involved in a monitoring program.

## APPENDIX 2-A

### ANALYSES: PRINCIPLES AND METHODS

#### Introduction

Fossil fuel contamination may have several sources including petroleum, petrochemicals, and combustion products of petroleum and coal. Each of these sources contributes characteristic compounds or mixtures of compounds. Analytical methods must be selected or developed to identify and measure these indicator molecules. A rationale for an initial list of parameters to measure and a section on analytical methodology are included in this appendix.

Petroleum contains a large suite of molecular types including saturated, straight-chain hydrocarbons (HCs) (n-alkanes); saturated, branched-chain HCs (alkanes); saturated ring HCs (cycloalkanes); and substituted and unsubstituted aromatic HCs. Petroleum also contains a variety of nitrogen-, sulfur-, and oxygen-containing compounds that have only occasionally served as pollution indicators in tissue. These molecular types have wide ranges of molecular weights and boiling points. The highly volatile compounds are usually not contaminants of organism tissue.

Making accurate HC measurements in bivalves is often a difficult task. Furthermore, it is difficult to determine the souces of HCs from the data. The NRC has suggested criteria for differentiating petroleum HCs from biogenic HCs that have been applied over the past several years. (Note that not all differences apply to all organisms, nor to all crude oils and refined products.)

1. Petroleum contains a much more complex mixture of hydrocarbons than do biogenic mixtures, with much greater ranges of molecular structure and weight.

2. Petroleum contains several homologous series with adjacent members usually present in nearly the same concentration. The approximate unity ratio for even- and odd-numbered alkanes is an example, as are the homologous series of $C_{10}$ to $C_{25}$ isoprenoid alkanes. Marine organisms have a strong predominance of odd-numbered $C_{15}$ through $C_{21}$ alkanes.

3. Petroleum contains more kinds of cycloalkanes and aromatic hydrocarbons. Also, the numerous alkyl-substituted ring compounds have not been reported in organisms. Examples are the series of mono-, di-, tri-, and tetramethyl benzenes and the mono-, di-, tri-, and tetramethyl naphthalenes.

4. Petroleum contains numerous naphthenoaromatic hydrocarbons that have not been reported in organisms. Petroleums also contain numerous hetero-compounds containing S, N and O, metals, and the heavy asphaltic compounds (NRC 1975).

One criterion added since the NRC report is the $^{14}C$ activity of isolated HCs. If the bulk of the HCs are fossil, then there should be little $^{14}C$ activity. This criterion has enjoyed some success; however, isolation of sufficient HCs from bivalves to provide the requisite 0.1 to 1 g carbon for $^{14}C$ activity counting is a major task. A second criterion established since the NRC report is the determination of the stereochemistry of key molecules. For example, pristane from biological sources is composed solely of the 6(R), 10(S) isomer, while in petroleum the alkane exists as a mixture of three possible isomers (Patience and others 1979).

The combustion products of petroleum and coal contain substantial amounts of aromatic HCs. Due to the high temperature of combustion, the ratio of alkyl aromatics to the parent aromatic compounds is much smaller in the combustion products than in petroleum. The aromatic HCs formed by combustion are rich in unsubstituted aromatics, and the aromatic rings may be linked together to form the polynuclear aromatic hydrocarbons (PAHs) that are now regarded as widely distributed pollutants (Laflamme and Hites 1978). Thus, it is possible to obtain some insight into the sources of aromatic HCs by determining the relative ratios of parent aromatic HCs and alkylated homologs.

We recognize that laboratories initiating environmental HC assays will be initially constrained in any fossil-fuel-compound research or monitoring program by lack of more sophisticated analytical instrumentation and/or trained analysts. Thus we describe several levels of sophistication of measurement techniques. The choice of the techniques to be used will depend on the monitoring or research goals of each program and for this reason we cannot set forth a "cookbook" approach to methodology.

## Extracting Samples

Samples of mussels should be shucked under appropriately clean conditions. If only a portion of the sample is to be analyzed, the tissue should be homogenized and the unanalyzed material stored at -10°C or colder. Wet tissues can be extracted by digestion under alkaline conditions, either with aqueous KOH, followed by partitioning of HCs into ethyl ether (Warner 1976), or with ethanol and KOH, followed by addition of water and partitioning of HCs into pentane, hexane, or iso-octane (Dunn 1976). Alternatively, the sample may be dried (by freeze drying or by gentle warming), ground or broken into small particles, and extracted with a HC solvent, using hexane or preferably methanol/benzene or methanol followed by benzene in a Soxhlet apparatus (Farrington and others 1976). Alkaline digestion procedures are generally faster and simpler than Soxhlet extraction procedures, and yield as good or better recoveries (Farrington and others 1976, Warner 1976, Dunn 1976). Care must be taken to assure that the apparatus used for freeze drying does not contaminate the sample with diffused organics from pump oil or from other components. This has been an intermittent problem noted in one of our laboratories (J.W. Farrington, Woods Hole Oceanographic Institution, personal communication, December 1978). Dynamic headspace sampling of the aqueous caustic tissue homogenate has also been applied to these types of analysis (Chesler and others 1978).

Separation and Isolation of Hydrocarbon Fractions

Various procedures of column, thin-layer, gel permeation, and high-pressure liquid chromatography have been applied to non-saponifiable extracts to isolate HCs. However, there is no unique method that separates and resolves adequately the entire range of molecular types and weights found in petroleum-contaminated organisms.

Gearing and others (1978) have recently compared thin-layer chromatography (TLC) and column chromatography separations of HCs and found compatible results in determining total aliphatic and total aromatic-olefinic HCs in sediment extracts. Care must be taken during TLC to avoid photo-oxidative and evaporative losses. Since many of these separations are sensitive to temperature, humidity, grade and size fraction of solid absorbent, and geometry of the column or thickness of the TLC plate absorbent, each laboratory must check separation efficiencies for themselves. The following procedure can be used only as guidelines for a starting point.

Column and TLC Chromatography Procedure

The following procedure is from the U.S. Mussel Watch Program (Farrington and others 1979):

A clean-up column 9 mm i.d. $\times$ 35 cm long with a 300 ml reservoir is packed with silica gel 100-200 mesh deactivated 5% with water, and alumina 200 mesh also deactivated 5% with water.

The column is filled with approximately 100 ml of hexane. Then 8 gm of silica gel is slowly poured in allowing the air to escape followed by tapping for 1-2 minutes to promote settling. Solvent is allowed to drip through to further promote settling. On top of this 8 gm of alumina is added and the column is again tapped to settle the alumina into the column. All of the packing should reside in the column tube. The solvent is allowed to drip through until it just reaches the top of the packing.

Then the extract is loaded on top of the column and allowed to drip through until it reaches the top of the column packing. Four 1 ml rinses of the samples flask are then added to the column,

allowing each to wash through to the top of the
packing before adding the next rinse. Then another
15 ml of hexane is added to the column and allowed
to drip through. All of this hexane is collected
as fraction 1. Then 20 ml of 10% toluene, 90%
hexane and 20 ml of 20% toluene, 80% hexane are
added to the column separately and collected as
fraction 2. Finally, 20 ml of toluene is added to
the column and collected as fraction 3. These
column extracts are then concentrated down for
analysis.

Warner (1976) used a slightly different technique:

The column used was a 0.9 x 25 cm Fischer and
Porter glass column fitted with a fritted glass
disc, a Teflon stopcock, a 100 ml reservoir, and
Teflon seals. The column was packed by filling it
with petroleum ether and slowly adding 10.0 g of
silica gel (MCB No. SX-144-7, activated at 150°C
overnight), while vibrating the column gently with
an electric vibrator to remove bubbles. The
stopcock was opened and 2-3 psi of nitrogen
pressure was applied (using Teflon tubing and oil-
free fittings) until the solvent level was about 1
mm above the silica gel. The column was never
permitted to run dry; a thin layer of solvent was
left on top of the silica gel at all times during
its use. Any traces of hydrocarbon in the silica
gel were removed by washing the column with 25 ml
of methylene chloride followed by two 2 ml
petroleum ether rinses and a 40 ml petroleum ether
rinse prior to the addition of a tissue extract.
The elution rate was 1-2 ml/min.
    The concentrated tissue was transferred to the
column and allowed to move down the column using
nitrogen pressure until the solvent level was about
1 mm above the silica gel. The walls of the column
were rinsed with petroleum ether and nitrogen
pressure was applied until the solvent level was
again about 1 mm above the silica gel. The column
was then eluted with 25 ml of petroleum ether. The
eluate was collected in a 25 ml concentrator tube.
The total eluate at this point, fraction 1,
contained all of the saturated hydrocarbons. After
adding 50 ml of 20% methylene chloride in petroleum
ether, v/v, to the reservoir, two 25 ml eluates

53

were collected, fraction 2 and fraction 3.
Fraction 2 contained most of the mono- and
diaromatic hydrocarbons as well as most of the
biogenic olefinic hydrocarbons.  Fraction 3
contained most of the triaromatic hydrocarbons.

Quinn and Wade (1974) report a technique for TLC
separations as follows:

Plates were coated with a 0.37 mm layer of
silica gel G Type 60 (Analabs) containing no
organic binder, activated at 110°C for twelve hours
and then precleaned with $CH_2Cl_2:CH_3OH$ (80:20, v/v).
After spotting, samples were eluted by hexane with
1% $NH_4OH$ added to aid in visualization of bands.  A
mixture of 3-methylnonadecane and phenanthrene
standards was eluted on the same plate to act as a
guide in removing the hydrocarbon fractions of the
sample.  Bands were visualized by spraying the
plate with bromothymol blue and viewing under UV
light.

## Additional Sample Purification Steps

The aliphatic fraction from chromatographic separation
is usually analyzed directly by gravimetric, gas
chromatographic, and GC/MS techniques.  Further separation
of n-alkanes and branched alkanes from cycloalkanes can be
achieved, e.g., with urea clathration (Wade and Quinn 1979)
or 5-A molecular sieves.  Additional cleanup is often needed
for the "aromatic" fractions as they often contain
substantial amounts of olefinic material.  This has been
accomplished by various procedures.  Giger and Blumer (1974)
and Giger and Schaffner (1978) used LH-20 Sephadex to
further isolate the aromatic HCs.  Dunn and co-workers (Dunn
and Stich 1976) used selective partitioning of aromatic HCs
into dimethyl sulfoxide (DMSO).  The basis of selection of
this procedure has recently been extensively studied
(Natusch and Tomkins 1978).  Recent work by Dunn (1979)
suggests that LH-20 and DMSO approaches are complementary
and remove different interfering compounds.  High-pressure
liquid chromatography (HPLC) is another approach to
obtaining better isolation of polycyclic aromatics from
olefinic compounds (Warner and others 1979, Wise and others
1977).

## Quantitation and Further Analysis
## of Hydrocarbon Fractions

### Gravimetric Determination

The aliphatic and aromatic fractions from the column chromatographic separation are made up to volume in a volatile solvent. Aliquots are spread in thin layers in small pans on an ultramicro balance (sensitivity 0.1 $\mu g$) and evaporated to determine the residue weight, as a direct measure of the amount of HC in each sample. The procedure is simple and rapid, but requires standardization of procedures to avoid under- or over-evaporation of the solutions. The procedure is not as sensitive as gas chromatographic measurements but is well-suited to studies where it is not necessary to have information on the chemical nature of the components of the fractions. Gravimetric analysis is easily combined with other analytical procedures.

### Determination of Aromatic Hydrocarbons by Fluorimetry

HCs are extracted from mussels and are either subjected to partial cleanup to remove lipids (Zitko 1975) or to a normal aliphatic-aromatic fractionation. The extract is made up to volume and its fluorescence measured in a scanning fluorimeter. Excitation and emission wavelengths are scanned with either the excitation wavelength held constant and the emission wavelength altered (Zitko 1975, Eaton and Zitko 1979), or both wavelengths scanned simultaneously, with the excitation wavelength a constant amount less than the emission wavelength (Gordon and Keizer 1974, Popi and others 1975). Examples are given in Figures 2-A.1 and 2-A.2. Repetitive emission scans at a number of different excitation wavelengths can be used to construct a fluorescence contour diagram showing intensity at all combinations of excitation and emission wavelengths (Hargrave and Phillips 1975). The fluorescence of mussel extracts is compared with that derived from standard oils, both with respect to the shape of the spectrum that can be used for source characterization or identification, and the intensity of fluorescence (for quantitation).

The procedures are relatively simple. However, the method has severe limitations as a quantitative technique. Often the source of contamination of mussels is unknown, and frequently more than one source is involved. This may make

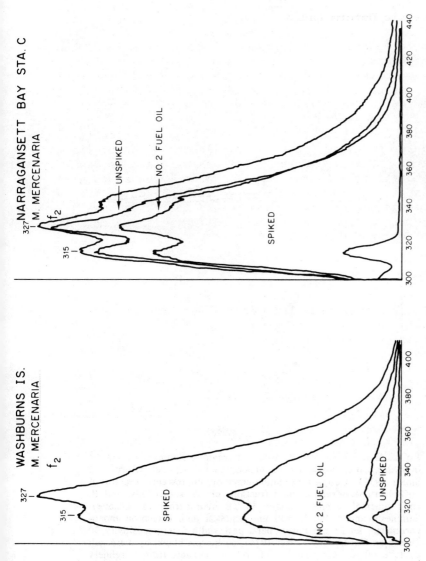

FIGURE 2-A.1  Ultraviolet fluorescence spectra of a 2- and 3-ring aromatic hydrocarbon fraction isolated from clams (*Mercenaria mercenaria*) and clams "spiked" with No. 2 fuel oil (Farrington and others 1976).

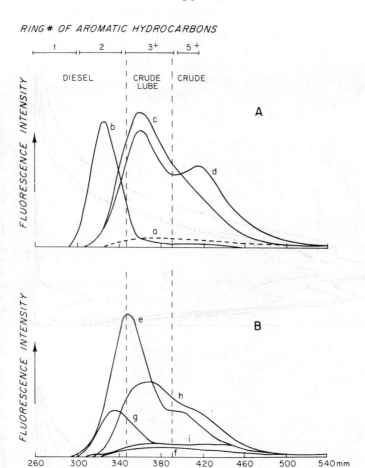

FIGURE 2-A.2 Synchronous-scan ultraviolet fluorescence spectra of hydrocarbons isolated from mussels and fossil fuels (Burns and Smith 1977). Coupled excitation-emission fluorescence spectra of 20% benzene/hexane column fractions of: A, source oils; and B, environmental samples (about 50 μg hydrocarbon/ml hexane). Samples scanned 260 to 540 nm emission with excitation monochrometers set 23 nm lower. Band pass width 5 nm; sensitivity 30 (based on Gordon and Keizer 1974). a) hexane blank; b) diesel oil; c) lube oil; d) Gippsland crude oil; e) mussels from a refinery wharf; f) clean mussels; g) mussels from small boat wharf; h) sediments near refinery; i) clean sediments.

it difficult, if not impossible, to choose an oil with which
to standardize the fluorimeter. Mussels may contain
background fluorescence of nonpetroleum origin. Also,
weathering and/or selective bioaccumulation effects may
radically alter the fluorescence characteristics of aromatic
residues in mussels, relative to the original source oil or
oils. However, if the source of the aromatic HCs in tissues
is known, the ease and speed with which fluorescence
measurements may be made may make the technique very useful
for investigating hot spots of contamination. Thus Eaton
and Zitko (1979) have used fluorescence measurements
effectively in analyzing over 800 samples in an
investigation of environmental creosote contamination.

Thin-Layer Chromatography Followed by
Fluorescence Detection

     For measurement of higher-molecular-weight, polycyclic
aromatic HCs, there is an extensive literature on thin-layer
chromatographic separations of isomeric PAHs. After
chromatography, compounds are measured fluorometrically
either in situ, or after elution from the adsorbent.
Chromatography on cellulose acetate is a specialized
technique that has the ability to separate the important
carcinogen benzo(a)pyrene (BaP) from all other PAHs,
including the isomeric benzo(e)pyrene (a noncarcinogen).
This capability has allowed the development of techniques
for monitoring BaP in mussels, which use this compound as an
indicator of high-molecular-weight carcinogenic PAHs (Dunn
1979; Dunn and Stich 1975, 1976; Dunn and Young 1976).
Although limited to the measurement of only one compound,
the procedures are reliable and precise (standard deviation
of replicate analyses less than 10 percent). BaP monitoring
has been used to investigate the contamination of coastal
waters by creosote (Dunn 1979; Dunn and Stich 1975, 1976;
Dunn and Young 1976), and to investigate baseline levels of
PAH contamination on the coast of southern California (Dunn
and Young 1976) and the coast of Oregon (Mix and others
1977).

High-Pressure Liquid Chromatography

     Recent developments in the field of HPLC suggest that in
the near future this technique will be a major method for
the analysis of aromatic HCs.

HPLC uses high-pressure pumps to force eluting solvent through columns of packing with small particle sizes (10 to 5 $\mu$m). The very small size of the packing and the high porous regularity increases the resolution of the columns markedly over classical TLC or column chromatography. Compounds eluting from the column are detected by their ultraviolet (UV) absorption or their fluorescence in special low, dead-volume detectors. Both methods of detection may be used in series. Since the detection is nondestructive, the eluate from the column may be collected so that peaks may be analyzed by other techniques such as MS. Columns are easily scaled up for preparative work. Longitudinal diffusion in chromatography columns is very slow, so it is possible to stop the flow of eluting solvent while peaks are in the detector, and to do stopped-flow UV absorption or fluorescence spectrum scans. Detection limits are on the order of 0.1 to 1 ng for UV absorption, and at least an order of magnitude better for fluorescence detection. As both UV and fluorescence detectors respond with widely different sensitivities to different compounds and change their response factor as wavelength settings are altered, it is essential to have reference chemicals for each compound that is to be quantitated. If it is desired to measure one compound to the exclusion of another interfering compound, it is often possible to select either adsorption or fluorescence wavelengths to achieve this. The ratio in response between UV and fluorescence detectors can be used to identify or confirm the identity of peaks. If two compounds are present with the same peak, it can be used to determine the concentration of each HC through simultaneous equations.

Microparticulate columns packed with silica can be used with HC solvents to fractionate aromatic HCs according to classical normal-phase techniques. Although resolution is higher than with TLC, many PAH isomers of interest are still difficult to separate. The most promising approach to the analysis of aromatic HCs from environmental samples involves the use of microparticulate reversed-phase columns that are packed with silica whose surface is coated with a chemically bonded organic phase. Solvents for this system usually consist of water with different concentrations of acetonitrile or methanol, or more generally gradients with increasing amounts of organic solvent. Aromatic compounds generally elute in order of increasing lipophilic nature. Compounds containing more condensed rings elute later than lower-molecular-weight compounds, and alkylated derivatives of PAHs elute later than their parent non-alkylated HC.

A large number of commercial reversed-phase column packings are available, and although most are based on the same support and derivatizing reagent, there are distinct differences in selectivity between different brands. The best separations of 4-, 5-, and 6-ring PAHs have been on octadecylsilic (ODS) packings (Vydac TP reversed, and Perkin Elmer HC-ODS). Figure 2-A.3 shows the resolution of 5 isomeric, 4-ring PAHs by chromatography on Vydac reversed-phase packing. Figure 2-A.4 shows the resolution of 18 PAHs on the same column. The separation of the various 4-, 5-, and 6-ring PAHs is of particular significance, as these compounds are difficult to separate by GC, and the measurement of individual compounds is desirable as some are carcinogens.

Aside from providing a technique for the analysis of the major parent PAH compounds, HPLC shows considerable promise for determining the degree of alkylation of aromatic compounds. Extracts of mussels contaminated with aromatic HCs from creosote or combustion sources ("pyrolytic"-type contamination) contain mainly parent PAH compounds, which are fairly well resolved on available columns that have approximately 2,500 theoretical plates. However, aromatics from petroleum sources show a more confused chromatogram, with alkylated derivatives of PAHs of a given ring size co-eluting with the parent compound of the next larger ring size. In contrast, an aminosilane liquid chromatographic packing has been shown to fractionate aromatic HCs mainly according to the number of aromatic rings, with little regard as to the degree of alkylation of the rings (Wise and others 1977). This column can be used as a preanalytical fractionation step to produce fractions containing mainly one ring size, but having different degrees of alkylation. The fractions can then be analyzed on reversed-phase columns to determine the relative levels of the parent PAH compounds and their various alkylated derivatives. This approach is very promising for the analysis of samples containing a substantial proportion of alkylated aromatic compounds.


Measurements of Hydrocarbons by Gas Chromatography

The aliphatic HC fraction can be analyzed by GC using flame ionization detection. In the past both packed columns, SCOT(R) columns, and capillary columns were used. The liquid phases were generally nonpolar, such as SE-52, SE-30, OV-101, Apiezon L, or Dexsil 300. We strongly recommend the use of higher resolution glass capillary GC as

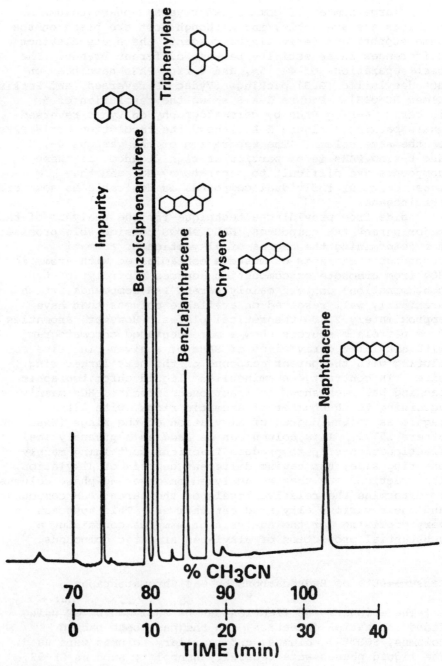

FIGURE 2-A.3  Reversed-phase HPLC separations of isomeric 4-ring PAHs (Wise and others 1977).

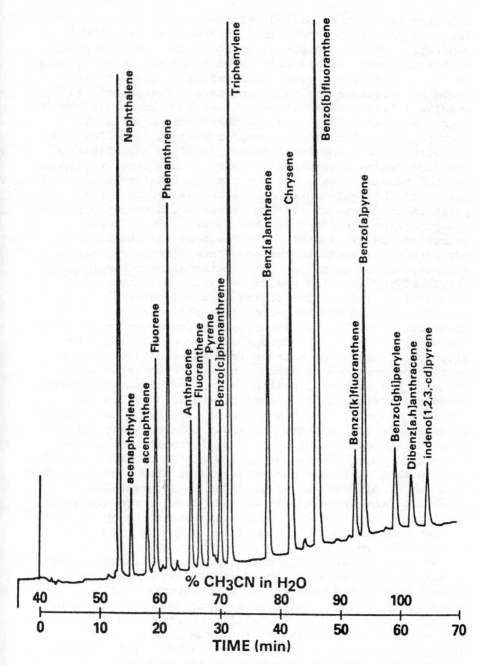

FIGURE 2-A.4    Reversed-phase HPLC separations of PAHs on the U.S. Environmental Protection Agency's Priority Pollutant List (Wise and others 1977).

a better choice than packed columns or SCOT columns. The improved separations and added information gained by separation of more compounds overcome the need for greater technical expertise and care in handling the glass capillary columns. High-resolution glass capillary GC is rapidly becoming a routine analysis procedure. The procedures resolve a large number of individual compounds as peaks, but there is often a large hump of unresolved material in chromatograms ("unresolved complex mixture," or UCM) from polluted areas. The total amount of HC in the aliphatic fraction can be determined by summing the detector response for the resolved peaks and for the UCM. Information on the nature of the contaminant oil can be gained from the position and shape of the UCM and from the resolved peaks. A distinction can often be made between biogenic and petroleum HCs by examining the ratio of the abundance of odd-carbon n-alkanes to the abundance of even-carbon n-alkanes, and by examining the relationships between pristane, phytane, and n-alkane levels (Farrington and others 1976). Some idea of the original nature of the oil (heavy or light) can be gained from the molecular weight distribution of the HCs found both as resolved peaks and in the UCM. Figure 2-A.5 shows a gas chromatogram of the aliphatic fraction of tissues from mussels taken from a harbor contaminated by chronic pollution compared with a chromatogram of the aliphatic fraction from mussels sampled immediately after a No. 2 fuel oil spill. Due to the problems associated with most small laboratory integration systems, special software is needed to accurately integrate resolved GC peaks, eluting on top of complex unresolved mixtures. Three-dimensional plots of the distribution of HC components of interest can be generated from digital GC data using an XY plotter, as illustrated in Figure 2-A.6. Generally, the above described analytical procedure gives recoveries for the higher-molecular-weight HCs ($n$-$C_{15}$ to $n$-$C_{31}$ and naphthalene to benzopyrene) of greater than 85 percent and has a precision of $\pm$ 5 percent.

As with the aliphatic fraction, GC with flame ionization detection resolves a number of individual aromatic HCs, but often leaves a UCM hump in the chromatogram. If the aromatic HC fraction has not been purified other than by chromatography on silica or silica alumina, the chromatogram will often contain peaks arising from olefins and unidentified biogenic HCs (Figure 2-A.7). If the source of the aromatic fraction is predominantly petroleum, the chromatogram will contain a large number of peaks arising from alkylated PAHs, whereas if the source is coal tar or

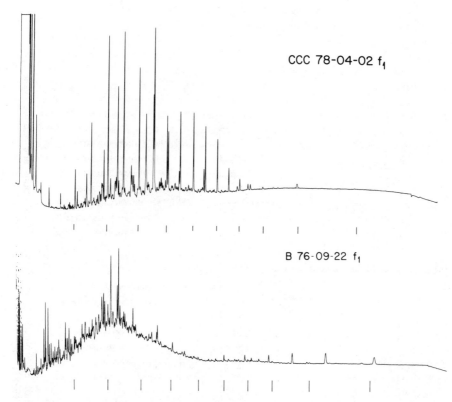

CCC 78-04-02 f₁

B 76-09-22 f₁

FIGURE 2-A.5 Glass capillary gas chromatograms of hydrocarbons from mussels (*Mytilus edulis*) (U.S. Mussel Watch Program, unpublished data, 1978). *Top:* Mussels contaminated with No. 2 fuel oil sampled one day after the spill. *Bottom:* Mussels from Boston Harbor, Massachusetts, U.S.A. Both are alkanecycloalkane fractions isolated from mussels. Tick marks at bottom of gas chromatograms indicate the elution positions of even carbon *n*-alkanes from $C_{12}$ to $C_{30}$.

creosote or pyrolytic in nature (e.g., air pollution), the alkylated derivatives will be minor components in the chromatogram. This fact can be useful in assigning the origin of the tissue contamination. Figure 2-A.8 shows the gas chromatogram of the aromatic fraction from mussel tissue contaminated by a marine petroleum spill. Note the multiple alkylated derivatives of the major parent PAHs, which greatly complicate the chromatogram. Figure 2-A.9 shows the gas chromatogram of aromatics derived from a fossil fuel combustion source and, in addition, a "pyrolytic" pattern of compounds rich in the parent molecular structures, with relatively low concentrations of alkylated derivatives.

64

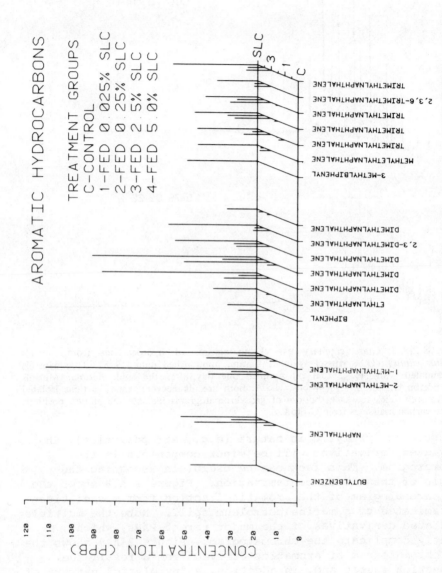

FIGURE 2-A.6 Three-dimensional computer plot of aromatic hydrocarbon concentrations and sample types (U.S. Mussel Watch Program, unpublished data, 1978).

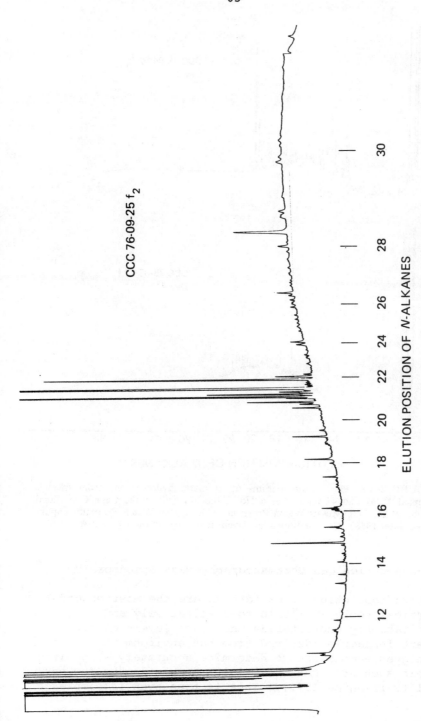

CCC 76-09-25 f₂

ELUTION POSITION OF N-ALKANES

FIGURE 2-A.7  Glass capillary gas chromatogram of aromatic-olefinic hydrocarbon fraction from mussels (*Mytilus edulis*) sampled in the Cape Cod Canal, Massachusetts, U.S.A. (U.S. Mussel Watch Program, unpublished data, 1978). Three most intense peaks are olefins of unknown structure. See legend to Figure 2-A.5 for GC conditions.

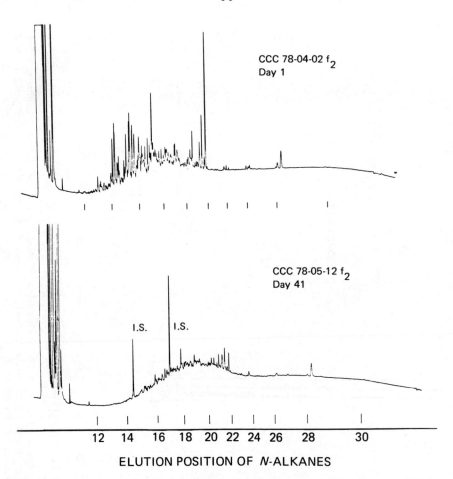

FIGURE 2-A.8 Gas chromatograms of aromatic hydrocarbons from mussels sampled 1 day and 41 days after a No. 2 fuel oil spill at the Cape Cod Canal station of the U.S. Mussel Watch Program (U.S. Mussel Watch Program, unpublished data, 1978). GC conditions are given in the legend of Figure 2-A.5.

## High-Resolution Gas Chromatography-Mass Spectrometry

High-resolution GC/MS methods are the most powerful analytical tools available to qualitatively and quantitatively characterize the trace levels of organics present in samples derived from the environment. Such techniques enable one to determine accurately a specific organic such as a PAH in a tissue sample in the low ppb level (Farrington 1979). This is true even in complex

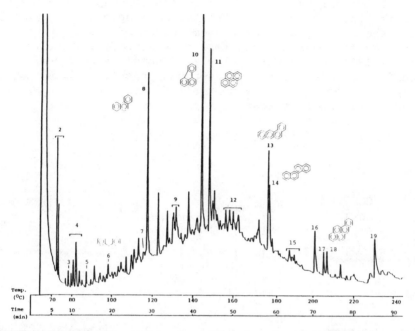

FIGURE 2-A.9  Gas chromatograms of aromatic hydrocarbons from mussels sampled near San Francisco Bay (South), California, U.S.A. (U.S. Mussel Watch Program, unpublished data, 1978). Column chromatography—Benzene eluate. Peak identification by GC/MS: 1. naphthalene; 2. $C_1$-naphthalenes; 3. biphenyl; 4. $C_2$-naphthalenes; 5. acenaphthene; 6. fluorene; 7. dibenzothiophene; 8. phenanthrene; 9. $C_1$-phenanthrenes; 10. fluoranthene; 11. pyrene; 12. $C_1$-fluoranthenes + $C_1$-pyrenes; 13. benz[a]anthracene; 14. chrysene; 15. $C_1$-benz[a]anthracenes + $C_1$-chrysenes; 16. unknown (M+ = 252); 17. benzo[e]pyrene; 18. benzo[a]pyrene; 19. cholesterol.

fractions isolated from mussel tissues that contain many other organics.

Chromatographically resolved components should be identified by more than simple retention data. This is accomplished by use of a GC/MS computer system. Eluting GC peaks are introduced directly into the ion source through a direct-type coupling to eliminate any sample loss in transfer from the GC to the MS system. The ionization potential is operated at 70 eV and inlet resolution in order to obtain typical low-resolution spectral data for interpretation. Scanning rates are typically 2 to 3 seconds/decade or less so that adequate information is available on each narrow GC peak. Because of the large quantities of data generated by such techniques, it is necessary to have a computer system to collect and store the data. Data display should include total ion summation plots

68

(total ion chromatograms), mass chromatography of selected masses, and background subtracted mass spectra. Figure 2-A.10 shows a typical mass chromatogram of a sample analyzed for PAHs from naphthalene (m/e = 128) to $C_5$-naphthalene (m/e = 198); and several alkylated homologs of these PAHs. Identification of specific compounds should be based on comparison of background subtracted mass spectra with mass spectra of known standards.

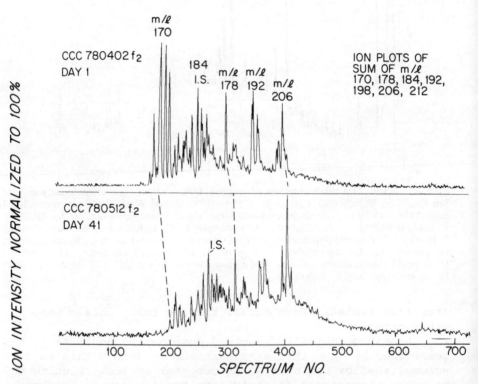

FIGURE 2-A.10   GC mass spectrometer computer systems analysis of aromatic hydrocarbons in *Mytilus edulis* from the Cape Cod Canal Station of the U.S. Mussel Watch Program after a No. 2 fuel oil spill (U.S. Mussel Watch Program, unpublished data, 1978). Analyses were accomplished using a Finnigan 1015 quadrupole mass spectrometer coupled to a Varian gas chromatograph equipped with a splitless injector. A 17 m by 0.32 mm ID SE-52 glass capillary column was installed in the system and operated as follows: Injection was at 40°C and 30 seconds in splitless mode, then rapidly brought to 70°C, and programmed to 240°C at 4°/min. Helium carrier gas was at 14 psi. The mass spectrometer operating conditions were 20 eV, 500 μm, scan from 40-500 atomic mass units (amu) every 3 seconds. Ion currents were integrated and plotted using the Riber 400 software package for the 1015 data system. m/e 170: C-3 naphthalenes; 184: C-4 naphthalenes, and dibenzothiophene; 192: C-1 phenanthrenes (and anthracenes, traces only); 206: C-2 phenanthrenes (and anthracenes, traces only). Internal standard: hexaethyl benzene.

Quantitative MS is recommended for the PAH components. Such methods require the use of a GC/MS computer system operated in a very clean and stable mode, as response curves must be constructed using a quantitative solution of aromatic HC standards. An integrated ion current plot derived from mass chromatogram data is then used to compare the known with unknown aromatics. Table 2-A.1 lists indicator aromatics that have been measured in the U.S. Mussel Watch Program.

## High-Resolution Gas Chromatography-Mass Spectrometry of Alkanes and Cycloalkanes

Fossil fuels contain a wealth of other compounds of biogeochemical significance (biological markers) that can be advantageously used as pollutant markers and, especially, in assessing chronic pollution. Steranes, methylsteranes, triterpanes, long-chain acyclic isoprenoids and monoaromatic steranes (Figures 2.1 and 2.2) belong to this category of compounds. Their ubiquitous presence in geological samples, namely in fossil fuels, and their geochemical stability and resistance to the biological metabolism contribute to making those series a valuable diagnostic tool. Both molecular and stereoisomeric distributions of the several members of the

TABLE 2-A.1  Most Prevalent Compounds Found in Fuel and Crude Oil[a] and in Combustion Source Emissions[b]

|  | m/e |
|---|---|
| Naphthalene | 128 (50% recovery) |
| Methylnaphthalenes | 142 |
| C-2 Naphthalenes | 156 |
| C-3 Naphthalenes | 170 |
| Phenanthrene (anthracene) | 178 |
| Methylphenanthrenes | 192 |
| C-2 Phenanthrenes | 206 |
| Fluoranthene | 202 |
| Pyrene | 202 |
| Chrysene/Benzanthracene/Triphenylene | 228 |
| Benzopyrene/Perylene/Benzofluoranthene | 252 |
| Dibenzothiophene | 184 |
| Methyldibenzothiophene | 198 |

[a]Naphthalenes, phenanthrenes, dibenzothiophenes.

[b]Phenanthrene, fluoranthene, pyrene and m/e 252 compounds.

series have to be considered.  Although they are generally
included in the unresolved hump of the HCs they can be
analyzed by mass fragmentography (Albaiges and Albrecht
1979).

The most abundant series in fossil fuels is that of the
triterpanes of hopane type (Figure 2.1).  This family is
formed by a series of $C_{27}$-$C_{35}$ members (Ensminger and others
1975).  The stereochemistry of the $C_{17}$ and $C_{21}$ in the
precursor biological materials are $\beta(H)$ with one
diastereomer at position 22, while in petroleum in matured
HC samples the $17\alpha(H)$, $21\beta(H)$, $21\alpha(H)$ with the 22R + 22S
isomers are found.  In addition, two $C_{27}$ members are
present, the $17\alpha(H)$ and $18\alpha(H)$-trisnorhopanes, their
relative abundance depending on the maturation conditions of
the product (Seifert and Moldowan 1978).

In the sterane and methylsterane families (Figure 2.1),
two series of compounds can be expected: the normal and the
rearranged steranes, the latter again as a result of the
geochemical maturation.  Variations in the stereochemistry
of carbons 5, 14, 20, and 24 afford complementary evidence
of fossil fuel origin, because the chiral centers are built
in only one stereochemical configuration by biosynthesis.

Long-chain cyclic isoprenoid HCs ($C_{25}$-$C_{40}$), although not
as ubiquitous as the preceding series, can be of interest in
the identification of specific sources of HC pollutants.
Even the stereochemical elucidation of the pristane should
allow distinction between a biological and a pollutant
source.  Mono-aromatized steranes (Figure 2.2) have recently
been found in geological samples and can also be proposed as
indicators of fossil fuel pollution.

Important differences in the relative distribution of
the various members of the series occur depending on the
origin and geochemical history of the sample.  However,
since they are exclusively geological maturation products,
their presence in contemporary environments is good evidence
of a fossil-fuel-derived input.  An example of a use of this
technique applied to whole oil samples is given in Figure 2-
A.11.  In principle, there is no reason why this would not
be applicable to analysis of fossil-fuel polluted bivalves,
although extensive applications have not been reported at
this time (Risebrough and others 1979).

In conclusion, it can be said that novel molecular
fingerprinting techniques, involving specific geochemical
markers, are expected to be more conclusive than gross
compositional parameters (total HC, UCM) for determining
sources of fossil fuel contamination.  High-resolution GC/MS
computer systems are able to furnish the corresponding

profiles without complex sample treatments.  Multiparametric profiles can be obtained from one run and can be easily stored for further processing.

### U.S. Environmental Protection Agency "Priority Pollutants"

Some of the organic compounds on the list of "Priority Pollutants" (PAHs, polychlorinated biphenyls, and chlorinated pesticides) shown in Table 2.1 are already being measured in mussel watch programs.  Analyses of the other organic compounds on the priority pollutant list have not been performed in bivalves in more than a few samples, if any.

The volatile organics can probably be analyzed via a modified headspace or vapor stripping technique as described in Chesler and others (1978).  The other nonvolatile organics can most likely be analyzed by a combination of solvent extraction, concentration, HPLC, high-resolution GC, and high-resolution GC/MS, as previously described for HCs. Careful verification of analyses will be needed before one can be certain that reliable qualitative and quantitative methodologies are available for these compounds. Furthermore, analyses should not be limited to only the "priority pollutants."  There are thousands of petrochemicals released in various parts of the world's coastal areas.  The list given in Table 2.1 is only a first step and includes some of the most ubiquitous compounds and those of the most grave concern from the point of view of general protection of public health and natural resource populations.

### Notes

1. From Burns and Smith (1977) and Smith and Burns (1978a, 1978b).
2. From Dunn and Stich (1975), Dunn and Young (1976), and Mix and others (1977).

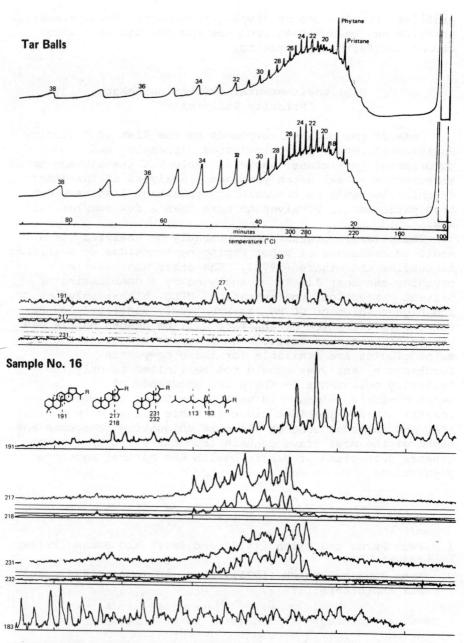

FIGURE 2-A.11  Gas chromatograms of tar ball and mass fragment search for steranes, triterpanes, and isoprenoid alkanes in tar balls and crude oil (Albaiges and Albrecht 1979). *Top:* Gas chromatograms of tar ball samples and fragment search for mass numbers 191, 217, and 231 for triterpanes and steranes. *Bottom:* Mass fragment search for triterpanes, steranes, and isoprenoid alkanes in Venezuela Laguna crude oil.

73

REFERENCES

Albaiges, J. and P. Albrecht (1979) Fingerprinting marine
pollutant hydrocarbons by computerized gas
chromatography-mass spectrometry. Int. J. Environ. Anal.
Chem. (in press).

Anderson, J.W. (1975) Laboratory Studies on the Effects of
Oil on Marine Organisms: An Overview. Publication No.
4249. Washington, D.C.: American Petroleum Institute.

Burns, K.A. and J.L. Smith (1977) Distribution of petroleum
hydrocarbons in Westernport Bay (Australia): Results of
chronic low level inputs. In Fate and Effects of
Petroleum Hydrocarbons in Marine Organisms and
Ecosystems, edited by D.A. Wolfe. New York: Pergamon
Press.

Burns, K.A. and J.L. Smith (1978) Biological monitoring of
ambient water quality: The case for the use of mussels
as indicators of certain organic pollutants. Marine
Chemistry Unit, Ministry for Conservation, 7B Parliament
Place, Melbourne, Victoria 3002, Australia (unpublished
manuscript).

Chesler, S.N., B.H. Gump, H.S. Hertz, W.E. May, and S.A.
Wise (1978) Determination of trace level hydrocarbons in
marine biota. Anal. Chem. 50:805-810.

DiSalvo, L.H., H.E. Guard, and L. Hunter (1975) Tissue
hydrocarbon burden of mussels as potential monitor of
environmental hydrocarbon insult. Environ. Sci. Technol.
9:247-251.

Dunn, B.P. (1976) Techniques for determination of
benzo(a)pyrene in marine organisms and sediments.
Environ. Sci. Technol. 10:1018-1021.

Dunn, B.P. (1979) Benzo(a)pyrene in the marine environment:
Analytical techniques and results. In Proceedings of the
International Symposium on the Analysis of Hydrocarbons
and Halogenated Hydrocarbons in the Aquatic Environment.
Hamilton, Ontario, Canada: Plenum Publishing Corp. (in
press).

Dunn, B.P. and H.F. Stich (1975) The use of mussels in
estimating benzo(a)pyrene contamination of the marine
environment. Proc. Soc. Exper. Biol. Med. 150:49-51.

Dunn, B.P. and H.F. Stich (1976) Monitoring procedures for
chemical carcinogens in coastal waters. J. Fish. Res.
Bd. Canada 33:2040-2046.

Dunn, B.P. and D.R. Young (1976) Baseline levels of
benzo(a)pyrene in southern California mussels. Mar.
Poll. Bull. 7:231-234.

74

Eaton, P. and V. Zitko (1979) Polycyclic aromatic
hydrocarbons in marine sediment and shellfish near
creosoted wharf structures in eastern Canada.
Charlottenlund, Denmark: International Council for the
Exploration of the Sea (in press).
Ensminger, A., A. Van Dorsselaer, C. Spyckerelle, P.
Albrecht, and G. Ourisson (1975) Pentacyclic triterpenes
of the Hopane type as ubiquitous geochemical markers:
origin and significance. Pages 245-260, Advances in
Organic Geochemistry 1973, edited by B. Tissot and F.
Bienner. Paris: Editions Technip.
Farrington, J.W. (1979) An overview of the biogeochemistry
of fossil fuel hydrocarbons in the marine environment.
In Marine/Aquatic Environment: Symposium on Analytical
Chemistry of Petroleum Hydrocarbons. Washington, D.C.:
American Chemical Society (in press).
Farrington, J.W., J.M. Teal, and P.L. Parker (1976)
Petroleum hydrocarbons. Chapter 1, Strategies for Marine
Pollution Monitoring, edited by E.D. Goldberg. New York:
Wiley Interscience.
Farrington, J.W., A. Davis, N.M. Frew, and E.D. Goldberg
(1979) Hydrocarbons in mussels and oysters of the U.S.
East and Gulf Coasts, 1976 (in preparation).
Fossato, V.U. (1975) Elimination of hydrocarbons by mussels.
Mar. Poll. Bull. 6:7-10.
Fossato, V.U. and W.J. Canzonier (1976) Hydrocarbon uptake
and loss by the mussel Mytilus edulis. Mar. Biol.
36:243-250.
Gearing, J.N., P.J. Gearing, T.F. Lytle, and J.S. Lytle
(1978) Comparison of thin layer and column
chromatography for separation of sedimentary
hydrocarbons. Anal. Chem. (in press).
Giger, W. and M. Blumer (1974) Polycyclic aromatic
hydrocarbons in the environment: Isolation and
characterization by chromatography, visible, ultraviolet
and mass spectrometry. Anal. Chem. 46:1663-1671.
Giger, W. and C. Schaffner (1978) Determination of PAH in
the environment by glass capillary gas chromatography.
Anal. Chem. 50:243-249.
Gilfillan, E. and J. Vandermeulen (1978) Alterations in
growth and physiology in chronically oiled soft shell
clams, Mya arenaria, chronically oiled with Bunker C
from Chedabucto Bay, Nova Scotia, 1970-1976. J. Fish.
Res. Bd. Canada 35:630-636.
Goldberg, E.D., V.T. Bowen, J.W. Farrington, G.R. Harvey,
J.H. Martin, P.L. Parker, R.W. Risebrough, W. Robertson,

75

E. Schneider, and E. Gamble (1978) The mussel watch.
Environ. Conserv. 5:101-125.
Gordon, D.C. and P.D. Keizer (1974) Estimation of petroleum
hydrocarbons in seawater by fluorescence spectroscopy:
Improved sampling and analytical methods. Fish Res. Bd.
Canada Tech. Rept. 481:29.
Hargrave, B.T. and G.A. Phillips (1975) Estimates of oil in
aquatic sediments by fluorescence spectroscopy. Environ.
Pollut. 8:193-215.
Keith, R.H. and W.A. Telliard (1979) Environmental Science &
Technology Special Report: Priority pollutants. I. A
perspective view. Environ. Sci. Technol. 13(4):416-423.
Kidder, G.M. (ed.) (1977) Pollutant Levels in Bivalves: A
Data Bibliography. Available from the U.S. Environmental
Protection Agency, Environmental Research Laboratory,
South Ferry Rd., Narragansett, Rhode Island 02882.
Laflamme, R.E. and R.A. Hites (1978) The global distribution
of polycyclic aromatic hydrocarbons in recent sediments.
Geochim. Cosmochim. Acta 42:289-303.
Mix, M.C., R.T. Riley, K.I. King, S.R. Trenholm, and R.L.
Schaffer (1977) Chemical carcinogens in the marine
environment: Benzo(a)pyrene in economically important
bivalve molluscs from Oregon estuaries. Pages 421-431,
Fates and Effects of Petroleum Hydrocarbons in Marine
Organisms and Ecosystems, edited by D.A. Wolfe. New
York:  Pergamon Press.
National Research Council (1975) Petroleum in the Marine
Environment. Washington, D.C.: National Academy of
Sciences.
National Research Council (1976) Assessing Potential Ocean
Pollutants. Washington, D.C.: National Academy of
Sciences.
National Oceanographic and Atmospheric Administration (1978)
Ocean Pollution Research, Development and Monitoring
Needs. Report of a Workshop at Estes Park, Colorado,
July 10-14, 1978. Boulder, Colo.: NOAA, Environmental
Research Laboratory.
Natusch, D.F.S. and B.A. Tomkins (1978) Isolation of
polycyclic organic compounds by solvent extraction with
dimethyl sulfoxide. Anal. Chem. 50:1429-1434.
Patience, R.L., S.J. Rowland, and J.R. Maxwell (1979) The
effect of maturation on the configuration of pristane in
sediments and petroleum. Geochim. Cosmochim. Acta 43 (in
press).
Phelps, D.K. and W.B. Galloway (1979) A Report on the
Coastal Environmental Assessment Station (CEAS) Program.
Paper No. 13, presented to the International Council for

76

the Exploration of the Sea Symposium/Workshop on
Monitoring of Biological Effects of Pollution on the
Sea. In Rapports et Procesverbaux des Reunions, Conseil
International pour l'exploration belamer Charlottenlund
slot Denmark, edited by A. McIntyre and J. Pearce (in
preparation).

Popi, M., M. Stejskal, and J.H. Mostechy (1975)
Determination of polycyclic aromatic hydrocarbons in
white petroleum products. Anal. Chem. 47:1947-1950.

Quinn, J.G. and T.L. Wade (1974) Marine Memorandum Series
No. 33. Kingston, R.I.: University of Rhode Island.

Risebrough, R.W., B.W. deLappe, W. Walker II, A.M. Springer,
M. Firestone-Gillis, J. Lane, W. Sistek, E.F. Letterman,
J.C. Shropshire, R. Wick, and A.S. Newton (1979) Pattern
of hydrocarbon contamination in California coastal
waters. In Proceedings, International Congress on
Analytical Techniques in Environmental Chemistry, edited
by J. Albaiges. London: Pergammon Press (in press).

Seifert, W.K. and J.M. Moldowan (1978) Applications of
steranes, terpanes and monoaromatics to the maturation
and source of crude oils. Geochim. Cosmochim. Acta
42:77-95.

Smith, J.L. and K. Burns (1978a) Hydrocarbons in Westernport
Bay Mussels. Final Report, Hydrocarbons in Victorian
Ecosystems, Task T01-702. Westernport Regional
Laboratory and Marine Chemistry Unit, Ministry for
Conservation, Australia.

Smith, J.L. and K. Burns (1978b) Hydrocarbons in Port
Phillip Bay mussels. Final Report, Hydrocarbons in
Victorian Ecosystems, Task T01-707. Westernport Regional
Laboratory and Marine Chemistry Unit, Ministry for
Conservation, Australia.

Stegeman, J. and J.M. Teal (1973) Accumulation, release and
retention of petroleum hydrocarbons by the oyster,
Crassostrea virginica. Mar. Biol. 22:37-44.

Vandermeulen, J. and D.C. Gordon, Jr. (1976) Re-entry of
five year old stranded Bunker C fuel oil from a low
energy beach into water, sediments, and biota of
Chedabucto Bay, Nova Scotia. J. Fish Res. Bd. Canada
33:2002-2010.

Vandermeulen, J. and W.R. Penrose (1978) Absence of aryl
hydrocarbon hydroxylase (AHH) in three marine bivalves.
J. Fish. Res. Bd. Canada 35:643-647.

Wade, T.L. and J.G. Quinn (1979) Incorporation,
distribution, and fate of saturated petroleum
hydrocarbons in sediments from a controlled marine
ecosystem. Mar. Envir. Res. (in press).

Warner, J.S. (1976) Determination of aliphatic and aromatic hydrocarbons in marine organisms. Anal. Chem. 48:578-583.

Warner, J.S., R.M. Riggin, and A.P. Graffeo (1979) Recent advances in the determination of petroleum hydrocarbons in zooplankton and macrofauna. Symposium on Analytical Chemistry of Petroleum Hydrocarbons in the Marine/Aquatic Environment. ACS Symposium Series. Washington, D.C.: American Chemical Society (in press).

Wise, S.A., S.N. Chesler, H.S. Hertz, L.R. Hilpert, and W.E. May (1977) Chemically-bonded aminosilane stationary phase for the high performance liquid chromatographic separation of polynuclear aromatic compounds. Anal. Chem. 49:2306-2310.

Wise, S.A., S.N. Chesler, H.S. Hertz, W.E. May, F.R. Guenther, and L.R. Hilpert (1979) Determination of trace level hydrocarbons in marine biota. Pages 41-52, Proceedings of the International Congress on Analytical Techniques on Environmental Chemistry, Barcelona, Spain, Nov. 27-30, 1978. Paper submitted through the National Bureau of Standards. London: Pergamon Press (in press).

Zitko, V. (1975) Aromatic hydrocarbons in aquatic fauna. Bull. Environ. Contam. Toxicol. 14:621-631.

# 3

TRACE METALS


DAVID J.H. PHILLIPS (Chairman), Fisheries Research Station,
    Hong Kong
TURGUT BALKAS, Middle East Technical University, Icel,
    Turkey
ANTONIO BALLESTER, Instituto de Investigaciones Pesqueras,
    Barcelona, Spain
KATHE K. BERTINE, San Diego State University, San Diego,
    California
CHARLES R. BOYDEN, Portobello Marine Laboratory, Dunesin,
    New Zealand
MARKO BRANICA, Center for Marine Research, Zagreb,
    Yugoslavia
DANIEL COSSA, Institut National de la Recherche
    Scientifique, Quebec, Canada
DAVID H. DALE, The Papua New Guinea University of
    Technology, Lae Papua, New Guinea
RAFAEL ESTABLIER, Instituto Investigaciones Pesqueras,
    Caddiz, Spain
TSU-CHANG HUNG, Institute of Oceanography, Taipei, Taiwan
JOHN MARTIN, Moss Landing Marine Laboratory, Moss Landing,
    California
MICHAEL J. ORREN, University of Capetown, Capetown, South
    Africa
DONALD K. PHELPS, Environmental Protection Agency,
    Narragansett, Rhode Island
JOHN E. PORTMANN, Ministry of Agriculture, Fisheries and
    Food, Essex, Great Britain
JOAQUIN ROS, Instituto Espanol de Oceanografia, San Pedro
    del Pinatar, Spain
JACK F. UTHE, Halifax Laboratory, Nova Scotia, Canada
DAVID YOUNG, Southern California Coastal Water Research
    Project, El Segundo, California

## THE CONCEPT OF INDICATOR SPECIES

Studies on the trace metal content of different water
bodies may be performed using samples of water, sediments,
or members of the indigenous biota.  Many researchers
currently favor the use of biota.  The species selected are
termed "indicator" or "sentinel" organisms.  Quantitative
estimates of the relative degree of contamination of
different areas by trace metals may theoretically be made by
measuring the accumulation of the elements within the
tissues of the organisms.

The use of indicator organisms has certain advantages
over that of water or sediments in defining the trace metal
abundance in a study area.  The organisms provide a record
of trace metal accumulation integrated over a period of
months, eliminating the need for frequent sampling (as in
water analysis).  In addition, the use of organisms
circumvents the need for assumptions about the biological
availability of trace metals necessary in other methods of
study.  There are, however, disadvantages in the use of
indicator organisms.  Among them are the introduction of
biological variables and the differences in the availability
of metals to different species, which may lead to
uncertainty in the interpretation of results from indicator
surveys.

## METHODS OF ANALYSIS OF MATERIAL

### Basic Techniques

A variety of analytical methods for the processing of
biological samples and subsequent determination of their
trace metal content have been developed and are described in
the literature.  In capable hands, most of these methods can
yield reliable results.  The members of the panel were not,
therefore, disposed to endorse the use of any one analytical
technique; it was felt that the choice of technique was one
that should be based on individual circumstances rather than
any generalized subjective opinion.  Most experienced
researchers use their own preferred methodology and probably
would produce more consistent and reliable results with
their own methods than they would if asked to change to a
set universal technique.  Several manuals exist that might
be used as sources of reference methods if such were needed
by either experienced or inexperienced personnel (Topping
and Holden 1978, IDOE 1973, FAO 1974).  Other key references

that may be of some use can be found in the reference
section of this chapter.

Atomic absorption spectrophotometry is probably the most
widely used technique for the analysis of trace metals in
biota, requiring relatively little in the way of sample
preparation or sophisticated instrumentation.
Notwithstanding the relative simplicity of this technique, a
thorough understanding of the problems involved (e.g.,
effects of sample matrix, interference problems) is
necessary before reliable results may be obtained. At least
the same degree of experience is needed for any other method
currently in use to analyze biota for their trace metal
content. It should be noted that even if the use of atomic
absorption spectrophotometry were universal, differences in
sample preparation prior to analysis might in some cases
produce variations in the final results.

## Problems of Contamination or Metal Loss

Contamination of samples has been experienced at one
time or another by most or all researchers. Sources of
contamination are various and often take some time to
identify and explain.

Losses of trace metals during sample preparation or
handling is another serious issue. Loss of metals in
hemolymph during bivalve shucking was of general concern; a
scheme for shucking that eliminates such losses has been
devised and is discussed later in this report. Metal loss
during sample drying is also of concern. The panel felt
that for homogenized bulked samples (preferably prepared
with an ultrasonic cutter to avoid contamination), an
aliquot should be taken specifically for drying and should
not be analyzed further. This process avoids the problems
associated with possible metal addition to or loss from
samples during drying. However, if bivalves are to be
analyzed individually, the use of aliquots is obviously
impossible. In this case, special care should be taken to
minimize the possibility of contamination during drying.
The experience of researchers concerning losses of volatile
metal species was that significant losses of species such as
alkyl lead might occur even at room temperature. For most
work, however, drying of samples at 70 to 100°C for 24 hours
is not unreasonable; little or no loss of the major metal
species will occur with this regime. It should be noted
here that the measurement of loss of metal salts added as a
spike to check sample integrity at drying is in no way

valid; losses at such stages should be studied by
comparative methodology, since metal salts added as a spike
may act quite differently from metals in the sample itself.


## Metals to Be Studied

The elements for study in an international monitoring
survey of bivalves should be selected on the basis of:

1. the current state of knowledge concerning analytical
techniques suitable for each element;
2. the potential of each element for pollution of
coastal or open-ocean regions;
3. the potential dangers of elements to man, via
ingestion of marine products; and
4. the need for an indication of sediment contamination
of the bivalve sample.

On these bases, the elements that should be selected for
study, in approximate order of priority, are:  mercury,
cadmium, lead, copper, zinc, and silver.  If possible, or in
areas where contamination is suspected, chromium, nickel,
and cobalt should also be studied.  Arsenic should be
included in all surveys, although here the speciation of
arsenic in the organism should also be elucidated (see
subsequent discussion).  Analysis of samples for iron and
aluminum should be carried out routinely for all samples,
primarily to serve as a check on sediment contamination on
the bivalves used.  Thus a total of 12 elements might be
included in routine monitoring of bivalves.


## Metal Speciation

To date, most authors have determined _total_ metal
burdens or concentrations in marine products, whether these
are to be related to ecosystem contamination or to possible
dangers to public health.  However, the panel recognized
that the determination of total metal disregards the form of
the metal in the organism, which may be of paramount
importance not only toxicologically but also as a clue to
metal sources in some cases.

One of the most relevant examples for which we now
possess some information concerns arsenic.  Inorganic forms
of arsenic are highly toxic to mammals, including man.  Thus
the high concentrations of total arsenic found in many

species of molluscs and crustaceans and some species of finfish from oceans or coastal waters have recently been a cause of concern. Occasional levels of greater than 100 ppm wet weight have been reported, and concentrations of 10 to 50 ppm are common in many organisms. Elevated concentrations of total arsenic have been found not only in coastal biota, but also in organisms from relatively unpolluted areas. If this arsenic were all in an inorganic form, the risk to public health would be very great. However, toxicological studies have shown that much of the arsenic in marine biota occurs in an organic form which is water soluble and rapidly excreted by mammals in the urine. The mammalian toxicity of this form of arsenic appears almost insignificant in comparison to that of inorganic species of the element. Recent work (by Scott Fowler and others, much of which is still in press) suggests that organoarsenates are synthesized by phytoplankton and passed up the food chain, where trophic level amplification may occur. The production of organoarsenates by phytoplankton is presumably species dependent and this process (and the concentrations attained by secondary consumers) has little relation to pollution. Clearly, if arsenic is to be included in a monitoring survey, the study of total arsenic in bivalves has little meaning. Methodology to separate the different arsenic species found in biota should be developed, and monitoring surveys should include studies of the occurrence of each species of arsenic in bivalves rather than simply determining total arsenic levels.

To a certain extent, these comments may apply also to other elements selected for study. However, in the case of other elements our current knowledge is often much more fragmentary. Organic species of mercury are well known and universally acknowledged as environmentally and toxicologically important, but suggestions of lead methylation in the environment (e.g., Wong and others 1975) have not been followed up by studies of lead speciation in marine or coastal biota. It may be assumed that each form of an element differs in its toxicological effect and in the flux or cycling of the contaminant through the ecosystem. The development of the appropriate techniques to separate and identify different forms of elements is urgently needed. In routine monitoring of bivalves at present, we suggest that the determination of total metals is sufficient for all metals other than arsenic and mercury. However, studies of hot-spot areas or areas that support biota exhibiting high levels of metal would benefit from more detailed studies of metal speciation in all compartments of the ecosphere. The

acknowledged importance of organic species of arsenic and
mercury is considered sufficient to justify analyzing
routine samples separately for organic and inorganic
species.


## Intercalibration of Analytical Results

The intercalibration of results from different
laboratories is possibly the most vital part of any
international monitoring venture. Results cannot be pooled
and interpreted unless they are known with complete
confidence to be comparable with respect to the analytical
technique. The decision to avoid use of a set procedure
lends even greater importance to intercomparison exercises.

An intercalibration exercise requires a reference sample
of sufficient quantity and homogeneity. For an
international bivalve monitoring program, at least one
reference material is necessary, preferably of one of the
major species used in the monitoring program itself. Unless
sufficient funding were available specifically for the
development of this reference sample, a body such as the
U.S. National Bureau of Standards (NBS), with the necessary
expertise to develop and certify such a standard, should be
approached. The reference material should be a homogenate
of the whole soft parts of an oyster (preferably Crassostrea
gigas) or a mussel (preferably Mytilus edulis). The sample
taken to derive the material should not be too highly
polluted, preferably containing metals in the lower end of
the range of concentrations exhibited by the species in a
monitoring survey. The differences between the
concentrations of some metals in oysters and mussels (e.g.,
zinc, copper) argue for the development of two materials,
one for each species; alternatively, one polluted and one
unpolluted location might be sampled to yield two reference
materials of the same species. The quantity of material
involved would be large, since it would be used both for
initial intercomparison tests and for ongoing quality
control. Materials currently considered to be international
standards (e.g., Bowen's Kale or the NBS bovine liver
homogenate) are thought to be less appropriate than a
bivalve reference material because of the differences in
matrix. The limited quantities of bivalve material
currently available from organizations such as the
International Council for Exploration of the Seas (ICES) or
the International Atomic Energy Agency (IAEA) are

insufficient to support a large-scale international
monitoring effort.

It should be emphasized that the mere distribution of an
intercomparison reference sample is not sufficient in
itself.  For any intercomparison exercise to be a success,
there must be coordination by an experienced laboratory
whose staff is prepared to follow up all anomalous results
in other laboratories and provide advice on resolving the
difficulties encountered.  The two-way exchange of data and
the follow-up phase are vital to the success of any
intercomparison exercise.  In addition, it is important to·
have a sufficient lead-in period to allow the performance of
intercomparison exercises before monitoring samples are
handled.  The analysis of field samples prior to the
successful conclusion of intercalibration procedures can
never be justified.

Experience with the ICES intercalibration exercises has
shown that the distribution of a standard salt solution of
known metal concentration is an invaluable aid to successful
intercomparison on reference or standard materials.  We
therefore suggest that such standard solutions be used in
any future international monitoring exercise in the initial
intercalibration phase.

During the lead-in phase prior to analysis of actual
samples, a reasonable standard of accuracy on a reference
material is $\pm 20$ percent of the certified value for metal
concentration.  Obviously a laboratory would have to achieve
such a standard of accuracy for all metals to be monitored
in the real survey.  It was considered that many
laboratories with previous experience in the analysis of
biota for metals could achieve the necessary level of
accuracy within 6 to 12 months if sufficient funds were
provided.  Other laboratories, conceivably those with less
experienced personnel, could be included in sample handling
at a later date once intercalibration exercises were
satisfactorily concluded.


Centralization and Training of Personnel

Of the several countries and laboratories that might be
interested in participating in an international monitoring
survey, some might initially lack either the appropriate
expertise or the facilities, or both.  One possible solution
to this problem is centralization of the program, but it
seems unrealistic to envisage sample analysis being
undertaken only by central or key laboratories.  Among the

difficulties of centralizing the monitoring are those involved with sample shipment and with the lack of a suitable central laboratory willing to undertake the large numbers of analyses needed in a venture of this scale.

Alternatives to the use of a centralized analytical facility must therefore be considered. In theory, experienced analysts could travel to the laboratories of inexperienced participants to initiate laboratory programs. However, this is unrealistic, as even the experienced analysts would need some time to become familiar with new equipment. Alternatively, a team of experienced researchers could be used, operating from its own vessel, with its own equipment. This suggestion would appear to be ruled out by the difficulties involved in covering a sufficient area in a given time; in addition, certain political problems might be involved, especially those of access to coastal waters by a foreign vessel.

For any long-term effort, monitoring programs conducted in developing countries would require trained local personnel operating in their own custom-built laboratories. These laboratories should have:

1. The necessary power supply and equipment, and an adequate supply of spare parts adequate to maintain the instruments used. Servicing of equipment should be available on a regular basis.

2. Access to a supply of chemicals and reagents of suitable quality. Conceivably a central supply could be maintained for those areas that have difficulty in rapidly acquiring a supply of reagents of a reasonable quality.

3. Expertise in the use of analytical methods and sampling techniques that have been shown to produce reliable data, or that are specified by a coordinating body.

An effective training program will have a number of characteristics. It may be assumed in the following discussion that a completely inexperienced analyst enters the training program at the beginning of Stage 1; one of greater experience might enter at later stages.

Stage 1: This stage comprises a basic program of laboratory training in analytical methodology. The period of training would depend on the speed of learning of the trainee, but it would demand several weeks if not months.

Stage 2: The trainee returns to his or her own laboratory, possibly accompanied for a short time by an adviser. The trainee spends an (unspecified) period becoming familiar with the instrumentation, evaluating methodology, and solving local problems.

Stage 3: The trainee analyst participates in an ongoing intercomparison study. Data may be produced on a "blind" or even "double-blind" basis if deemed necessary, and should be submitted to the training laboratory. If a sufficient level of competence has been achieved, the trainee may proceed to collect and analyze samples for a certain area of the international monitoring study.

It must be stressed that all analysts must participate in the initial intercalibration phase prior to their inclusion as active participants in the international monitoring study. The degree of accuracy necessary for acceptance into the international program should be $\pm 20$ percent around the certified mean value.

## POSSIBLE SPECIES FOR USE AS TRACE METAL INDICATORS

The requirements for an ideal indicator species have been discussed elsewhere in the literature and will not be repeated here. For a species to be a candidate for international monitoring, it should not only satisfy all the requirements suggested for local surveys (Phillips 1977b, 1978b), but should also exhibit a wide-ranging international distribution.

Bivalve molluscs are the most likely candidates for international monitoring of trace metals. Considerable information is available on the response of these organisms to trace metals in their ambient environment, aiding interpretation of results from monitoring surveys. In addition, bivalves in general satisfy more of the basic indicator requirements than do other organisms. The only serious contenders for use in a large-scale monitoring survey would be macroalgae. However, no species are known that can match the geographical distribution of some of the bivalves. Bivalve molluscs are not ideal as worldwide indicators of trace metals in all respects; for example, no one species extends through both tropical and temperate waters. However, it is felt that selected species of bivalves are the best available alternative on which to base a large-scale monitoring program. Other types of biological indicators might be used in areas defined by preliminary bivalve surveys as those of high metal contamination; for example, in these so-called hot-spots, the use of other indicator types might aid delineation of metal speciation or availability.

Constraints on the choice of species of bivalve molluscs for use in a monitoring program are mainly those of the restricted distribution of many bivalve species. Certain problem areas exist; for example, the eastern Mediterranean Sea has few bivalves except for scattered isolated populations of _Mytilus galloprovincialis_. In such areas, transplantation may be a useful technique.

In most temperate waters of the world, however, mussels of the genus _Mytilus_ can be found. The most widespread (and most frequently used) species is _Mytilus edulis_, which may be found in waters of salinities 5 to 35°/oo and temperatures from close to 0°C up to about 26°C. Lower salinities and higher temperatures are lethal, and restrict the wide distribution of this mussel in areas such as the Baltic and Mediterranean Seas. In the latter, _M. galloprovincialis_ is present. On the west coast of the United States, _M. edulis_ and _M. californianus_ coexist, sometimes separated according to tidal habitat. These species of mytilids are similar to each other, and may exhibit similar metal kinetics. _Mytilus edulis_ is undoubtedly the bivalve of choice for an international monitoring survey in temperate waters; much is known not only of its metal kinetics and indicator abilities, but also of its physiology.

In the tropical and subtropical belts, no widespread species of _Mytilus_ occurs, although some regions support related mytilids of the genus _Septifer_. In such areas, oysters, especially the genus _Saccostrea_, have a superior distribution to that of mussels. _Saccostrea cuccullata_, for example, has been reported from locations as widespread as northwest Australia, the Philippines, East Africa, South Africa, India and Pakistan, and Hawaii. However, the speciation of these oysters is a matter of some uncertainty (Ahmed 1975) and, as this may affect the results of indicator surveys, of some importance. Oyster taxonomy is an area where further research is needed, especially in tropical regions.

Included among other possible candidates for international monitoring surveys are oysters of the genus _Crassostrea_, especially _C. gigas_ and _C. angulata_, which are often considered to be the same species in current literature. These oysters are widely distributed due to their culturing for food. They spawn in water temperatures higher than 21°C, and may therefore be found wild in waters of these temperatures. Some coasts contain _C. gigas_ that cannot spawn due to the temperature of the waters; in these cases, spat is continually introduced by aquaculturists, most commonly from Japan. Once again, speciation of these oysters is complex. Some authors contend that C. virginica and C. rhizophorae are at least closely related to C. gigas and C. angulata, as successful hybridization of the species has been noted. It should be understood that oyster nomenclature is somewhat confused here. For example, Saccostrea cuccullata is considered by Stenzel (1971) as a species identical to the commercially grown Crassostrea commercialis on the east coast of Australia. However, oysters of these groups, which are tuberculated or denticulated, are not closely related to the Crassostrea group mentioned above (C. gigas and others), as hybridization invariably fails. This report thus considers the Saccostrea and Crassostrea groups as separate genera (as does Stenzel). Within these genera, differences between species become indistinct, speciation occurring gradually within interbreeding populations.

In conclusion, the most likely candidates for use in an international monitoring survey are the mussel Mytilus edulis and relatives in temperate waters, and oysters of the genera Saccostrea and Crassostrea in tropical or temperate regions, or both. Studies of the metal contents of closely related species of oysters or mussels in areas of overlapping distribution would be most illuminating, not only as a guide to speciation but also as a way to investigate whether a constant ratio exists between the metal concentrations present in two species from similar habitats. More conventional taxonomic and distribution studies should also be encouraged, and data of this type would be a valuable spin-off from any attempt at worldwide monitoring studies.

## FACTORS AFFECTING RESULTS

Several factors may interfere with the interpretation of results from monitoring surveys for trace metals in bivalve indicators. Thus, for example, the concentrations of metals present may often be correlated with bivalve size or age. If there is a significant difference in size of bivalves sampled at different locations, the results will not be directly comparable in terms of environmental abundance of metals. Other factors that may affect trace metal concentrations in bivalves are season of sampling, sex of bivalve and its vertical position on the shoreline, and salinity and temperature of the ambient environment. These factors are considered here in turn, in order to produce a sampling method that will lead to reproducible sampling at all locations and eliminate the effects of variables as far as possible. Much of the following discussion centers on data from the literature for Mytilus edulis or Crassostrea gigas, as these are major candidates for use in international monitoring surveys. Fewer data are available on Saccostrea species, and further research should be performed on species of this genus to improve their use as accurate indicators of trace metals.

## Season of Collection

Seasonal changes in the concentrations and total body burdens of metals in bivalves have been noted for both mussels and oysters. Theoretically, sampling of all study sites at the same time of year should eliminate the effects of seasonal variation on the results of indicator surveys. However, this is not so for two basic reasons. First, simultaneous sampling of all sites is rarely practicable. Second, variables that are the basis of seasonal changes in the concentrations of trace metals in bivalves are not constant with location; e.g., water temperatures may vary between locations at any one time, as may the time of spawning in the bivalve populations.

Comparison of seasonal maxima and minima of concentrations of several trace metals in different species of bivalves leads to a general understanding of the possible magnitude of the effects of this parameter on the results of indicator surveys. Data for M. edulis have been published by Phillips (1976a), Goldberg and others (1978), and Majori and others (1978), and unpublished data are available from panel members Boyden and Cossa (Table 3.1). For zinc, cadmium, copper, iron, and manganese, the ratios of maximum to minimum concentrations through one year are generally a factor of three or less. There is some suggestion of location dependence in the data (see discussion by Phillips 1976a), in that mussels from open marine areas appear to exhibit less seasonal variation in trace metal concentrations than do those of estuarine areas. The observation may be related to the more marked changes in water quality that may occur in estuaries in periods of high runoff. Such an explanation may also be relevant in interpretation of observations of seasonal changes in metal concentrations present in other species of Mytilus. Thus

Goldberg and others (1978) could find little seasonal
fluctuation in the concentrations of several metals in M.
californianus collected from the open, marine waters off
Bodega Head, California, whereas Fowler and Oregioni (1976)
found location- and metal-dependent seasonal changes of
three- to ten-fold for M. galloprovincialis taken from
several locations in the northwestern Mediterranean Sea.

Reports of seasonal variation in the concentrations of
trace metals found in species of Crassostrea indicate that
these may be even more severe than those in mussels.  Table
3.2 shows that concentrations of cadmium, copper, and zinc
vary seasonally in C. virginica by factors of 2.3 to 5.4,
and in C. gigas by factors of 2.7 to 5.9.  As oysters are
often grown in estuaries, such greater differences between
seasonal maxima and minima might be expected, and appear to
confirm the above hypothesis.  Unfortunately, no information
on seasonal variability of trace metals in species of
Saccostrea is available.

The data show that seasonal variation in concentrations
of trace metals found in bivalves is severe, and may be
greatest in estuarine regions, where the seasonal
fluctuation in water quality is greater than that in open
oceans.  If sampling for monitoring surveys were to
inadvertently include bivalves from different locations that
were in different parts of the seasonal cycle, the
interference of these differences could be highly
significant.

The question is how the effects of seasonal variation
can be eliminated in samples taken in monitoring surveys
that cover a large area and use bivalve populations with
nonsynchronous seasonal profiles.  To answer the question,
some knowledge of reasons for seasonal variation in trace
metal levels in bivalves is useful.  Bryan (1973) suggested
that variation in trace metal levels in the scallops Pecten
maximus and Chlamys opercularis was related to phytoplankton
productivity; in periods of high productivity, metal
concentrations in phytoplankton fall, producing a
concomitant decrease in metal levels in scallops.  However,
the theory discounts changes in metal inputs from
terrigenous sources with season, and it is not generally
accepted.  Phillips (1976a) postulated that seasonal changes
in total body burden of metals in bivalves were minimal, and
that changes in concentration were thus reciprocal to
variations in body weight.  This leads to a seasonal maximum
in concentrations soon after spawning, when tissue weight is
at a minimum.  Data from Fowler and Oregioni (1976) for
Mytilus galloprovincialis support this hypothesis, as do

TABLE 3.1  Seasonal Maximum and Minimum Concentrations and the Ratio of the Two Concentrations in Populations of the Mussel *Mytilus edulis* from Marine and Estuarine Locations[a]

| Location | Data | Cadmium | Copper | Iron | Manganese | Zinc | Source |
|---|---|---|---|---|---|---|---|
| Port Phillip Bay, Australia | Maximum | 0.61 | — | — | — | 60.1 | Phillips (1976a)[b] |
| | Minimum | 0.20 | — | — | — | 19.3 | |
| | Ratio | 3.1 | — | — | — | 3.1 | |
| Narragansett Bay, USA | Maximum | 2.0 | 13.2 | — | — | 199 | Goldberg et al. (1978)[c] |
| | Minimum | 1.1 | 6.7 | — | — | 81 | |
| | Ratio | 1.8 | 2.0 | — | — | 2.5 | |
| Gulf of Trieste, Italy | Maximum | — | — | 50 | 15 | 30 | Majori et al. (1978)[b] |
| | Minimum | — | — | 20 | 10 | 10 | |
| | Ratio | — | — | 2.5 | 1.5 | 3.0 | |
| Helford Estuary, U.K. | Maximum | 2.4 | 17.4 | — | — | 225 | Boyden (unpublished data)[c] |
| | Minimum | 1.1 | 7.6 | — | — | 106 | |
| | Ratio | 2.2 | 2.3 | — | — | 2.1 | |
| St. Lawrence Estuary, Canada | Maximum | 0.60 | — | — | — | — | Cossa (unpublished data)[c] |
| | Minimum | 0.18 | — | — | — | — | |
| | Ratio | 3.3 | — | — | — | — | |

[a]All data refer to whole soft parts of mussels.
[b]Wet-weight basis for concentrations, $\mu$g/g.
[c]Dry-weight basis for concentrations, $\mu$g/g.

TABLE 3.2  Seasonal Maximum and Minimum Concentrations and the Ratio of the Two Concentrations in Whole Soft Parts of Oysters *Crassostrea virginica* and *Crassostrea gigas* from Three Locations ($\mu$g/g dry weight)

| Location | Species | Data | Cadmium | Copper | Zinc | Source |
|---|---|---|---|---|---|---|
| Rhode River, USA | *C. virginica* | Maximum | 16.0 | 280 | 5,100 | Frazier (1975) |
| | | Minimum | 3.0 | 60 | 2,200 | |
| | | Ratio | 5.3 | 4.7 | 2.3 | |
| Humboldt, USA | *C. gigas* | Maximum | — | — | 9,000 | Harrison (1978) |
| | | Minimum | — | — | 3,300 | |
| | | Ratio | — | — | 2.7 | |
| Helford Estuary, U.K. | *C. gigas* | Maximum | 4.5 | 700 | 5,100 | Boyden[a] (unpublished data) |
| | | Minimum | 1.5 | 120 | 900 | |
| | | Ratio | 3.0 | 5.8 | 5.7 | |

[a]Panel member.

those of Frazier (1975, 1976) and panel member Boyden
(unpublished data) for the oysters <u>Crassostrea</u> <u>virginica</u> and
<u>C. gigas</u>. In general, the seasonal maximum coincides with
the period of greatest run-off (in winter, in temperate
climates); thus metal availability from land-based sources
may also be greatest at this time. Phillips (1976a,b)
suggested on the basis of these data that the ideal time for
sampling mussels in a trace metal monitoring survey should
be spring, when concentrations of metals are at their
maximum and the differences between locations are also
greatest. Oysters, however, spawn in late spring and summer
(and, occasionally, in autumn) in temperate climates. In
tropical climates, oysters reach maturity very quickly
(sometimes in less than one month) and may spawn throughout
the year. Clearly, to avoid the effects of nonsynchronous
tissue weight changes (allied to spawning) in different
populations of bivalves, samples for monitoring surveys
should be taken during periods of least change in tissue
weight.

It is clear that if synchronous spawning of bivalve
populations is established, the correct time for sampling
would be just after spawning. However, in a monitoring
survey of international scale, nonsynchronous spawning of
populations would certainly occur, and the subsequent rapid
changes thereafter would have a significant effect on
seasonal fluctuation in metal concentrations of bivalves.
Thus surveys of bivalves should be conducted by sampling in
the winter months (a prespawned state for mussels, and
midway between spawning periods for oysters), corresponding
to that time when body burdens and concentrations, as well
as significant environmental and biological parameters, are
most constant. Although the absolute concentrations of
metals in mussels are generally lower during this period
than in late spring, the advantage of site-to-site synchrony
of mussel condition should reduce the interfering effects of
seasonal variability to an acceptable level. A further
advantage accrues from the need for prespawned mussels in
studies of biological effects; the single annual collection
envisaged for routine bivalve surveys of trace metals will
allow all samples to be collected at one period.


Size of Bivalves Sampled


As noted above, bivalve size or age often may be
correlated with the concentration of a metal present in the
whole soft parts. However, the rule is not general, and

significant size/concentration relationships are seen for
only some of the metals present in any given species.  In
addition, seasonal changes (again related to the sexual
state of bivalves) may affect the size/concentration
regression for any species/metal pair.

Possible effects of size on concentrations of metals in
bivalves sampled in monitoring surveys, and hence on the
interpretation of survey data in terms of the relative
contamination of areas by trace metals, can be eliminated by
sampling bivalves of identical size at each location.
However, in practice this is often impossible because of
differences in the dominant sizes of mussels in different
locations.  For this reason, most surveys are conducted
using a range of sizes.  The U.S. Mussel Watch Program
selected mussels of shell lengths from 50 to 80 mm from all
locations (Goldberg and others 1978).  This range was
considered in order to investigate the magnitude of residual
size-based differences in concentration that might be
expected.  For mussels (Mytilus edulis) ranging in shell
length from 50 to 80 mm, a corresponding dry tissue weight
range was found to be 0.52 to 3.00 g; for oysters
(Crassostrea gigas) of the same size range, the dry tissue
weight range was 0.17 to 3.08 g (panel member Boyden,
unpublished data).  If we consider the tissue weight ranges
to cover approximately one cycle of logs, and assume a
regression slope of -0.25 for the relation between (log)
metal concentration and (log) tissue weight (Boyden 1974,
1977), the possible residual difference due to size of
sampled individuals is about 100 percent, i.e., smaller
individuals might contain twice the concentration of larger
individuals within the size range studied.  Although this
degree of difference may never be attained (as selection of
bivalves at only one end of the size range in any one
location is unlikely), the magnitude of the difference seems
sufficient to justify attempts to further eliminate such an
effect.

Elimination of the extraneous effects of size is
theoretically possible by one of three alternatives.

1.  Study of only those species/metal pairs where size
has no effect on the concentrations of metals present.  In
practice, this is impossible to achieve, because it would
demand the sampling of too many bivalve species at each
location.  The alternative may therefore be discarded.

2.  Use of a weight-normalization procedure, designed
specifically to eliminate effects of size (tissue weight) on
metal concentrations found in bivalves.  Such a procedure

was used by Johnels and others (1967) in comparing mercury
concentrations in pike (<u>Esox</u> <u>lucius</u>) of different sizes from
different locations, and was adapted by Phillips (1976a)
for use with <u>Mytilus</u> <u>edulis</u>.  It is clear that if the
size/concentration regression exhibits a constant slope in
all populations of a given species at all times, the process
is an efficient and useful tool for eliminating size
effects; the regression slope, once known, can be applied to
all sample results, whether samples are analyzed using
individual animals or in bulk homogenate.  However, results
to date suggest that size/concentration regressions in
bivalves are not always of constant slope, but that they
vary between locations and occasionally within a single
location with season.  Unpublished data from panel members
Boyden and Cossa have been used to examine the regression
slopes relating size (tissue weight) of <u>Mytilus</u> <u>edulis</u> to
body burdens of cadmium, copper, and zinc present.  The data
for copper and zinc involve seven and eight regression
lines, respectively, for both English and Canadian mussel
populations.  A considerable degree of consistency is noted
here, slope values being $0.81 \pm 0.04$ and $0.84 \pm 0.03$,
respectively.  In these cases, therefore, limited data
suggest that the relation between bivalve tissue weights and
metal content is constant.  Thus the relation between tissue
weights and metal concentration is also constant, slope
values being $-0.19$ and $-0.16$ for copper and zinc,
respectively.  By contrast, data for thirteen regression
lines representing seven mussel populations in England and
Canada show cadmium contents relating to tissue weights of
$\underline{M}$. <u>edulis</u> with slope values of $0.81 \pm 0.23$.  The variability
of slope of the regressions is obviously much greater here,
and appears to occur with location at any one time and at
some locations with season.  In this case, weight-
normalization of results would be possible only if enough
individuals were analyzed at each location to fully
characterize the size/concentration regression for that site
and even with this large amount of work, the relation of the
cadmium levels in bivalves from each site would depend on
the weight used for the normalization process.  Analysis of
sufficient individuals (perhaps 40 per location) is
impractical for preliminary wide-ranging surveys, although
the method could be used in selected areas of interest in
follow-up surveys.  Thus weight-normalization methods,
although potentially useful in specific cases, are not
thought to be the answer to the need for eliminating
differences based on size in bivalve surveys.

3. Use of a limited size range. The final alternative appears to be the best means of decreasing the effects of size on the results of an international monitoring survey of bivalves. For mussels, a 5- or 10-mm range would be preferred instead of a 30-mm range of acceptable size, as in the U.S. Mussel Watch Program (50 to 80 mm). The narrower range places extra constraints on sampling in the field, but is nevertheless considered worthwhile.

Use of a limited range raises the issue of the appropriate size of bivalve to be used in monitoring surveys. A theoretical profile describing the relationship between metal concentrations and tissue weights of a bivalve is shown in Figure 3.1. The relationship may be similar to that found for copper in Mytilus edulis. Immature animals show an inverse relationship between tissue weight and concentration, and the relationship appears seasonally stable, i.e., slope values do not change with season. At maturity, concentrations may be unaffected by tissue weights or may decrease with increasing weight, according to season (and to site if seasonal profiles differ between sites). Furthermore, large individuals exhibit the quite different relationship of concentrations increasing with increasing tissue weight; it is not clear whether the latter relationship is seasonally stable or different from site to site, because data are incomplete.

Because of the constancy of the regression slopes, immature individuals merit consideration as being best suited to the requirements of international monitoring surveys. The use of immature animals has other advantages as well; the amount of material in bulk homogenates is significantly reduced, and the effects of bivalve sex and spawning are eliminated or at least significantly reduced. However, concentrations of some metals are highly dependent on size (tissue weight) in immature animals, and standardization of size is thus desirable. The shells of sampled mussels should be approximately 60 mm long; the maximum range of shell lengths permitted should be 50 to 80 mm. Experience in the U.S. Mussel Watch program and in other large-scale surveys indicates that mussels of this size are available at most sites. Oysters present more of a problem because less information relating size to concentration is available; however, species of Crassostrea and Saccostrea are known to exhibit size-dependent concentration changes, generally of negative slope, for some metals. A range of shell length slightly larger than that for the mussels would be acceptable for these species for

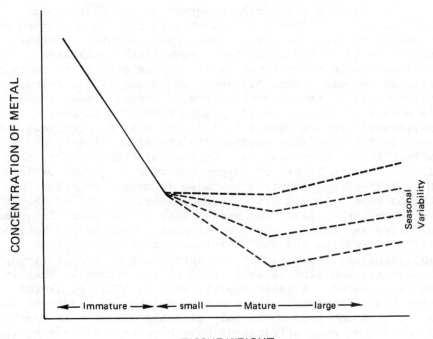

FIGURE 3.1 Theoretical profile for the relationship between metal concentration and tissue weights (whole soft parts) in a bivalve. (This profile is not intended as a general example, but as a specific example of extreme weight-dependence.)

use in an international monitoring survey. Individuals should be taken to produce mean shell lengths for the entire sample as close to 60 mm (mussels) or 75 mm (oysters) as possible.

### Sex of Bivalves Sampled

Since bivalves to be used in international monitoring surveys may be sexually immature in some cases, the effects of sex of the bivalves sampled on survey results may not be important. However, in some tropical areas, bivalves reach sexual maturity extremely quickly, sometimes in less than a month. In such cases, even small individuals may be sexually mature (Stenzel 1971).

Most bivalve monitoring programs do not distinguish between male and female individuals in samples collected for analysis of trace metals or other contaminants. However, some reports in the literature suggest that sex-based differences in trace metal levels may be present in certain bivalve species. Thus Alexander and Young (1976) found differences in the concentrations of copper, lead, and zinc in gonads of Mytilus californianus according to sex; male/female ratios were 0.5, 1.7, and 0.6, respectively. No such differences were evident for chromium, nickel, or silver in the same species. Gordon and others (1978) also found enhanced levels of copper and zinc in gonads of female M. californianus compared to those in males; male/female ratios were 0.36 and 0.17, respectively. Levels of cadmium, chromium, iron, lead, and nickel were no different in gonads of males and females in these samples. Watling and Watling (1976), studying the South African mussel Choromytilus meridionalis, found higher concentrations of copper, iron, manganese, and zinc in whole soft parts of females than in those of males. A later report (Watling 1978) suggested that the differences were almost two-fold for manganese and zinc, although copper was only slightly elevated in female mussels. No such differences have been reported for Mytilus edulis.

Differences in concentrations of trace metals exhibited by male and female individuals of the same species are probably caused by different biochemical characteristics of the gonad tissues in the two sexes. If so, the differences will be evident in whole soft parts only if gonad metal contributes a significant proportion of the total body burden of metals. Conceivably, such differences are also seasonally dependent, i.e., they vary according to gonad ripeness. Perhaps this explains the differences between the two sets of data for Mytilus californianus. Whatever the mechanism responsible for the difference, caution should be used in sampling mussels for a monitoring survey. Obviously such differences would be important only if some samples were predominantly of one sex. Further research is needed, especially on species such as Mytilus edulis and those of the genera Crassostrea and Saccostrea, designed to quantify sexual differences in pollutant levels and relate them to gonad chemistry and the sexual cycle. Such research might conceivably be performed during a lead-in phase to an international monitoring survey, concurrently with analytical intercomparisons, by a laboratory with recognized expertise. The direct analysis of gametes would also be of considerable interest, not only in this connection, but also to aid interpretation of possible effects of spawning on whole-body levels of trace metals in bivalves.

## Sampling Position on Shoreline

The exact vertical position on the shoreline at which bivalves are sampled is extremely important with respect to the metal levels encountered. For example, Nielsen (1974) found that concentrations of zinc, cadmium, lead, and iron in the cultured mussel Perna canaliculus varied with sampling depth at one location in New Zealand waters. It was suggested that this could be due to differences in available food or in the ratio of soluble to particulate metal in the water column with depth, or to the presence of hydrogen sulfide from sediment-dwelling bacteria affecting metal solubilities in the water column. Concentrations of the same metals in mussels from a second location, of greater water circulation, showed no vertical gradients of this kind. De Wolf (1975) reported higher concentrations of mercury at five different locations and at various times of the year in individuals of Mytilus edulis and M. galloprovincialis from the intertidal zone than in those sampled subtidally. Phillips (1976a) reported vertical gradients in concentrations of zinc, cadmium, and lead in M. edulis from a location situated approximately 1 km distant from a polluted freshwater outfall. The gradients were well defined in late winter, but were absent or much reduced in late summer. It was suggested that the gradients seen were produced by the exposure of mussels at the top of the water column to a metal-rich freshwater layer in winter conditions of high runoff from the adjacent catchment; in summer no such vertical stratification of the water column was present. Similar differences in the concentrations of several metals in mussels sampled at differing vertical shoreline positions were found by Stephenson and others (1978) in studies of mussels from the coast of California.

Such vertical gradients in the concentrations of trace metals observed in bivalves are presumably due to differences in metal availability with vertical position with respect to tides in most cases. It may be expected, therefore, that areas such as estuaries with marked salinity stratification would show a high degree of vertical metal changes in sessile mussels, whereas areas of greater vertical water circulation would show less emphasized profiles. It is most important to standardize the vertical collection position in monitoring surveys. Theoretically, the most reproducible sampling position both intrasite and intersite would be just below the spring low-water mark; however, the panel recognized that subtidal sampling was impractical and would create severe problems in some areas. Thus it was agreed that all samples for a large-scale monitoring program should be taken at the mid-tide level, halfway between low-water spring and high-water spring tidal marks. The vertical position of sampling should be carefully annotated for each location, and preferably also photographed, so that further sampling could be reproducibly undertaken if necessary. Extreme care should be taken in estuaries. Second- or third-phase studies in hot-spot areas (defined by the primary monitoring survey as contaminated locations) could employ sampling regimes involving the study of bivalves from different vertical positions.

## Inherent Variability

This report discusses several factors known to affect concentrations of trace metals in bivalves, such as season, size of animal, sex, and sampling position. However, even if all such variables are removed from a study, a certain amount of variation in metal concentrations in a bivalve population will be found. The residual differences are termed the "inherent variability" of the population, a variability that, basically, cannot be ascribed to any measured parameter.

The magnitude of inherent variability in a population may vary with a number of factors. Perhaps the most important of these are the particular species/metal pair considered and the level of contamination of the population as a whole. For example, cadmium levels in _Mytilus edulis_ typically exhibit less intrasite variability than those of zinc, lead, or copper. However, the variability of each of the metals increases markedly as the mean concentration of metal present increases, i.e., the standard deviation of the mean increases faster than the mean itself. This finding has led several authors to conclude that the use of parametric statistics to compare concentrations of metals in different bivalve populations may not be justified.

The inherent variability present in a population has been used to elucidate the minimum number of individuals needed to accurately characterize the average metal concentration found in a bivalve population. For this exercise, only species/metal pairs where metal concentration is known to be independent of tissue weight were considered, so as to eliminate the masking effects of size-dependent variation. Differences in sexual condition of individuals in any one population were minimized by the study of inherent variability at times other than those associated with spawning, thus ensuring stability of the seasonal component of intrasite variability. However, in the case of sex-based differences, no attempt was made to distinguish between the sexes of the individuals studied; thus the inherent variability here includes any sex-based differences in metal concentrations that might occur.

Figure 3.2 shows data from studies of Mytilus edulis in 1975 at Moss Landing, California. More recent data supplied by panel member J.H. Martin have suggested that the absolute levels of copper reported here may be somewhat high; however, the data are included to illustrate the approach used in studies of inherent variability. It can be seen from Figure 3.2 that a reliable estimate of the mean concentration of copper in the mussel population studied was produced only if the number of individuals studied was greater than 18. These data were also used to estimate the number of individuals needed to differentiate between two mussel populations differing in average copper concentration by 5, 10, 20, 50, or 100 percent (Figure 3.3). Using individual analysis (equivalent to one animal per pool), about four individuals from each site would have to be analyzed from each population to define a difference of 10 percent between population mean concentrations. The use of several individuals in each pool may be helpful in decreasing the numbers of analytical determinations made; e.g., with 10 individuals per pool, only about 20 pools (from each population) are necessary to distinguish a 10 percent difference between population means. It should be noted, however, that although such pooling decreases the number of analytical determinations from 200 to 40, the total number of mussels used in the latter example is double that of the former. Pooling of samples thus decreases the analytical workload, but increases the effort needed in sampling and preparation (shucking).

One of the panel members, C.R. Boyden, used a similar approach to study intrasite variability in the oysters Ostrea edulis and Crassostrea gigas. Forty individuals of each species were analyzed. Individual concentrations were then randomly extracted to generate 5 groups of 5, 5 of 10, 5 of 15, and so on up to 5 groups of 35 individuals. The variation between the coefficients of variation thus generated (Table 3.3) was studied to indicate the minimum sample size that sufficiently characterized the population.

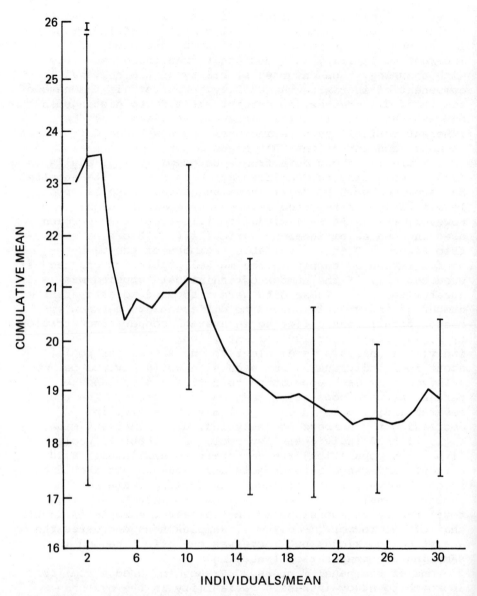

FIGURE 3.2 Cumulative means (and selected 95-percent confidence intervals) for the concentration of copper ($\mu$g/g dry weight) in 30 individual mussels (*Mytilus* species) (after Martin and Phelps 1979).

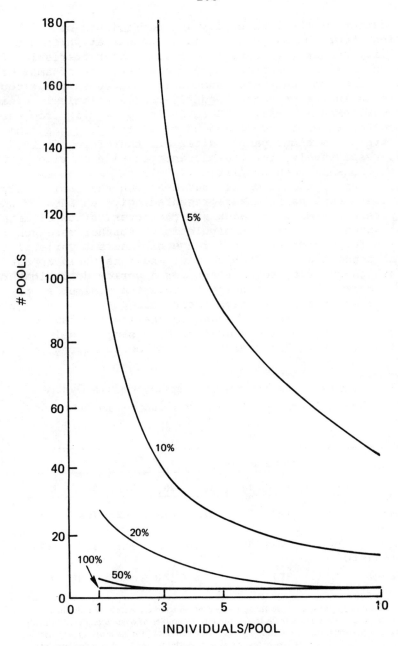

FIGURE 3.3  Plot showing the number of pooled mussel samples (ordinate) and the number of individuals per pool (abscissa) giving 90-percent certainty of detecting 5-, 10-, 20-, 50-, or 100-percent difference between means of copper concentrations at $P = 0.05$ using a $t$-test (after Martin and Phelps 1979).

Consideration of the variability in coefficients of variation shows "break-points," i.e., points at which the variability suddenly changes from a value characteristic of larger samples to a greater value characteristic of smaller samples. Thus the data show such points at subsample sizes of 10 to 15 in the case of O. edulis from a relatively clean environment (Menai Strait), 15 to 20 for O. edulis from the polluted Restronguet Creek, and 20 to 25 for C. gigas from Menai Strait. Minimum sample sizes are therefore 15, 20, and 25, respectively, for the characterization of zinc concentrations in each population.

Data such as these, while inadequate in that relatively few species/metal pairs are represented, give an idea of the size of sample needed for accurate characterization of metal concentrations in a bivalve population. Further research is needed to fully understand the range of inherent variability in metal concentrations in bivalves, and this would appear to be a high-priority requirement for a worldwide indicator survey. However, data at present suggest a minimum sample size of 25 individuals at each location.

Once a decision is reached as to the minimum sample size, the question of individual or bulk sample analysis must be addressed. The decision as to how to aggregate

TABLE 3.3   Variation of Zinc Concentrations (μg/g dry wright) in Oysters Expressed in Terms of Variability of Coefficients of Variation and Related to Sample Size [a]

| Species | Subsample Size | | | | | | |
|---|---|---|---|---|---|---|---|
|  | 5 | 10 | 15 | 20 | 25 | 30 | 35 |
| Ostrea edulis (Menai Strait) | 49.2 | 27.9 | 11.9 | 12.9 | 9.2 | 10.4 | 13.6 |
| Ostrea edulis (Restronguet Creek) | 49.3 | 30.4 | 24.1 | 18.6 | 20.3 | 19.5 | 18.7 |
| Crassostrea gigas (Menai Strait) | 31.7 | 29.5 | 16.0 | 18.0 | 8.8 | 9.0 | 8.3 |

[a] Means, standard deviations, and coefficients of variation were generated by random selection of individual concentrations from large population samples of oysters analysed individually for zinc concentration. Five series of values were generated for each oyster population. The coefficients for each sample size category were then determined and standard deviations obtained. Variability was then assessed by calculation of the coefficient of variation of these coefficients. Menai Strait is a "clean" environment; Restronguet is a zinc contaminated environment.

SOURCE: C. R. Boyden, panel member, unpublished data, 1978.

samples depends to a large extent on how accurate the
definition of differences between populations must be (see
Figure 3.3), as well as on practical aspects of the
availability of analytical time and personnel needed for
collection and shucking.  In terms of broad-based monitoring
surveys covering large areas, it was felt that initial
surveys should be done by creating a composite of all
individuals (25 or more) from each location and placing
extra capability into increasing the numbers of sites
studied.  Follow-up studies in areas exhibiting relative
enrichment of trace metals are envisaged as a second-tier
survey, where bulked samples may also be used for more
closely spaced locations to elucidate the spatial metal
distribution more accurately (i.e., to define the hot spot
area).  A third survey may then be deemed necessary, using
several locations around the hot spot (as well as a control
area) and analyzing bivalve samples individually rather than
in composites, to investigate possible effects of size and
other factors on the concentrations of metals found.  Other
indicator types like macroalgae may be useful in this phase
of study.  These aspects are discussed further in a
subsequent section.

## Salinity and Temperature

Parameters such as salinity and temperature are also
known to affect the uptake of trace metals by bivalves (see
Phillips 1976a, 1977a,c, 1978a).  These parameters are
important in that they may lead to difficulties in the
interpretation of results from monitoring surveys.  For
example, an increased concentration of cadmium in _Mytilus_
_edulis_ from hyposaline sites may be a function either of
greater absolute abundance of the element in those sites or
of greater biological availability of the element in areas
of low salinity.  It is therefore important to interpret
results from indicator surveys only with reference to the
availability of the element to the particular species being
monitored, in the first instance.
Because of the effects of salinity and temperature on
the uptake of metals by bivalves, a certain amount of
hydrological data is needed about study sites in monitoring
surveys.  The labile nature of these parameters suggests
that data should be amassed over a period of some months;
spot measurements of salinity and temperature during bivalve
sampling are not sufficient.  For many of the areas that
would be studied in an international monitoring survey, such

data already exist and are readily available. In other areas with no such information, steps should be taken to produce at least a general idea of these hydrological parameters. Furthermore, samples should not be taken in areas that are unusual with respect to either of these parameters (e.g., in heated effluents from power stations or other industries), because such samples would not be representative of overall water quality in that area.

## Depuration

Many researchers studying trace metal levels in marine or estuarine biota hold biota for some period in clean seawater immediately after collection in the field and prior to analysis. The aim of this depuration period is the complete excretion of gut contents, which may include digested and undigested food, and sediment particles, all of which may be rich in trace metals. Clearly, if depuration is not carried out, whole-body analytical results will include those metals associated with gut contents; this may contribute not only to individual variation in populations of biota, but may in some cases be a very significant contribution to the total metal load of the organism. The contribution of gut contents to measured total body loads or concentrations obviously depends on (a) the absolute concentrations of metals in the organism studied, (b) the nature and total amount of gut material, and (c) the concentration of gut materials relative to that of the species studied.

Bivalve molluscs are generally filter feeders or deposit feeders. Gut contents thus include inorganic particulates derived from the water column or from sediment, as well as planktonic food organisms. Both inorganic particulates and plankton are rich in trace metals; thus under some circumstances, the gut contents of bivalves may significantly affect the trace metal concentrations measured.

In an unpublished study, one of the panel members (C.R. Boyden) considered this problem in relation to copper in both mussels and oysters from clean and contaminated environments. The gut contents were assumed to contain only sediment and to contribute 2.5 percent of the total dry weight of the whole soft parts of each bivalve species. Calculations are shown in Table 3.4. It is noted that the sediment content of oysters might contribute only 1.0 to 2.5 percent of the total body load of copper, whereas that of mussels could represent 36 to 39 percent of the total body load of copper. The species difference here depends on the difference in the accumulation of copper by each species, oysters being much richer in copper than mussels. A similar result would be expected for the relative importance of zinc in the species, although the absolute importance of the gut contents in affecting bivalve zinc levels would be less than that for copper because both species contain higher absolute levels of zinc than of copper. By contrast, cadmium concentrations are similar in oysters and mussels; thus for cadmium, the percent contributed by gut contents to the total body load of metal would be similar for each species.

Data from the U.S. Mussel Watch Program and the California State Mussel Watch also include attempts to consider the importance of gut contents in affecting the total body loads or concentrations of metals in mussels (*Mytilus* *edulis* and *M*. *californianus*). Stephenson and others (1978) analyzed whole soft parts and the intestines of mussels separately. The mean ratio of iron in whole body to that in intestines was 0.06. It was suggested that this ratio could be used to normalize the other trace metals in whole body, thus eliminating the effect of gut contents on the analytical results. Calculations showed that the percentages of total body loads of each element contributed by the intestines were: Ag, 10 percent; Cd, 11 percent; Cr, 41 percent; Cu, 6 percent; Pb, 20 percent; Zn, 1 percent; Ni, 18 percent; Fe and Al, 100 percent; and Mn, 67 percent. Ouellette (1978) later produced results based on iron and aluminum normalization methods, using ratios calculated from observed depuration curves (see Figures 3.4 and 3.5). Using various assumptions, approximate percentage contributions of sediment in the gut to the total body load of metals in mussels were: Cr, 13 percent; Cu, 0.01 percent; Ni, 9 percent; Pb, 16 percent; and Zn, 0.007 percent. The disagreement between these values and the previously cited data reflects the different procedures used. However, it is clear that the concentrations of iron, aluminum, manganese, chromium, lead, and nickel in mussels may be significantly affected by gut contents.

Romeril (in press) measured the loss of elements from clams, *Mercenaria* *mercenaria*, during a holding period of several days in the laboratory. The data indicated a drop in zinc and copper concentrations of 9 and 12 percent, respectively, through the first 35 hours after field collection. Subsequent to this decrease, concentrations were constant, confirming the relatively long half-life of these elements in this species. Panel member Phillips

TABLE 3.4  Influence of Gut Sediment upon Whole Body Copper Content (µg) of the Mussel *Mytilus edulis* and the Oyster *Crassostrea gigas*[a]

| Copper Concentration of Sediment (<204 mµ) | Clean environment (Helford estuary, England) | Copper contaminated environment (Restronguet Creek, England) |
|---|---|---|
| | 160 µg/g | 2,000 µg/g |
| **Metal enrichment in *Mytilus edulis*** | | |
| Copper content of 1 g dry-weight mussel | 11 µg | 77 µg |
| Copper contained in gut bound sediment | 4 µg | 50 µg |
| Total copper content of 1 g mussel | 15 µg | 127 µg |
| % elevation due to sediment | 36% | 39% |
| **Metal enrichment in *Crassostrea gigas*** | | |
| Copper content of 1 g dry-weight oyster | 400 µg | 2,000 µg |
| Copper contained in gut bound sediment | 4 µg | 50 µg |
| Total copper content of 1 g oyster | 404 µg | 2,050 µg |
| % elevation due to sediment | 1% | 2.5% |

[a] Assuming gut content 2.5% of total body weight, thus 1 g dry-weight bivalve contains 25 mg of sediment.

SOURCE: C. R. Boyden, panel member, unpublished data, 1978.

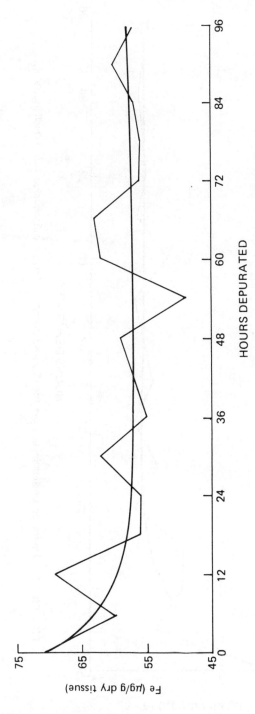

FIGURE 3.4  Depuration of iron from the digestive system of *Mytilus californianus* (after Ouellette 1978).

110

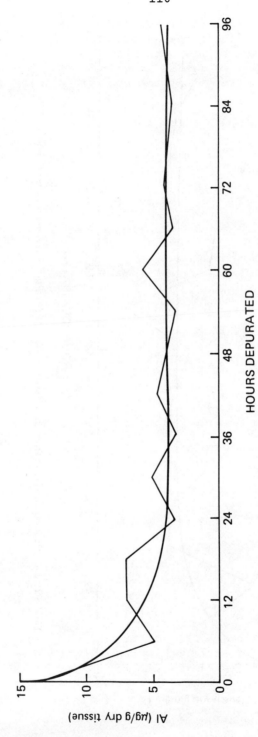

FIGURE 3.5  Depuration of aluminum from the digestive system of *Mytilus californianus* (after Ouellette 1978).

(unpublished data) has observed similar losses in mussels
(Mytilus edulis) kept in the laboratory.

It is important to note that the amount of perturbation
of total body loads of metals in bivalves by sediment or
other contents in the gut is not constant but varies
(especially with variation in the particulate content of the
ambient water). Thus, for example, sediment contents will
increase during periods of high runoff in estuaries or
storms in coastal waters. If these factors are different at
study locations, the effects of gut contents will vary
according to station.

The effects of gut contents on total body loads of
metals in bivalves are sufficient to justify the extra
effort and expense associated with depuration. Although the
use of a depuration period may lead to contamination
problems (e.g., by lead), it is generally felt that design
of an efficient noncontaminating process is possible and
that depuration of all samples should be attempted. The
differences in half-lives between pollutant types (e.g.,
metals and some HCs) would suggest that depuration should be
attempted only for stable metal and transuranic samples,
because there could be significant loss of organochlorines
and some petroleum HCs during depuration.


RELATED TOPICS

Weight Basis for Concentrations

Most authors currently report data for trace metal
concentrations on a dry weight basis, as this appears to
decrease somewhat the inherent variability component.
However, measurement of wet weights is not time-consuming,
and, in addition, most public health standards for metal
concentrations in seafood are based on wet weights rather
than dry weights. Results from an international monitoring
survey should be quoted on the basis of both wet and dry
weights.

Possible problems of contamination and metal loss during
drying have been referred to previously. For most work,
drying overnight at 70 to 105°C would not lead to
significant loss of metals. Where bulk homogenates are to
be analyzed, a separate aliquot of the homogenate should be
used to determine a ratio of dry weight to wet weight to
avoid possible contamination during this stage. Analysis of
individuals in later follow-up surveys of hot-spot areas
would not permit this, however, and these samples should be

dried prior to digestion and analysis, with special care taken during the drying process to avoid contamination.

Problems of metal losses during sample preparation should be studied during the lead-in phase of intercalibration by laboratories with recognized expertise. The problems are expected to be minimal, with the possible exception of arsenic, lead, mercury, and selenium. Silver is not considered to exhibit volatility problems, but the presence of lipids in the sample may interfere with correct determination of silver levels; this again should be investigated using the monitoring species themselves during a lead-in phase.

## The Use of Individual Tissues of Bivalves

There are advantages and difficulties associated with the use of individual tissues of bivalves or whole soft parts as indicators of environmental contamination. The use of individual tissues can be criticized for the following reasons:

1. the added logistical problems involved (e.g., dissection of tissues is necessary prior to sample freezing for storage);
2. the difficulty of dissecting tissues reproducibly from bivalves, encountered at least by relatively inexperienced researchers;
3. the dangers of metal loss in hemolymph and of cross-contamination of tissues during dissection; and
4. the significantly elevated cost of dissecting the necessary number of individuals for each composite to be studied.

For these reasons, it is felt that a general monitoring survey covering all locations should employ the whole soft parts of bivalves but should not routinely study individual tissues.

However, several advantages might accrue from the use of individual tissues of bivalves, especially when specific problems are to be addressed. For example, the fact that different tissues may exhibit widely different metal concentrations can be used to advantage when problems of analytical sensitivity are present. Data from many authors have shown that most or all bivalve species accumulate the highest concentrations of metals (including radionuclides) in the kidneys and digestive glands. These tissues might

therefore be used as a means of increasing the sample
concentration of metals and thus improve analytical
precision.  The use of the digestive gland as an indicator
organ has been pioneered by Young and coworkers (Alexander
and Young 1976, Young and others 1978).

A second advantage of specific organ analyses is that
different tissues have different response rates to changes
in exposure level.  For example, Eganhouse and Young (1978)
transplanted control zone mussels (Mytilus californianus) to
four depths (10 to 30 meters) on a taut-line buoy anchored
near a major submarine discharge of municipal wastewater off
Los Angeles, California (Royal Palms Beach).  Individuals
were collected from the four levels every 2 to 4 weeks for
half a year, and digestive gland, gonad, and adductor muscle
samples were analyzed for total mercury concentrations.
Resultant mean and standard error values are illustrated in
Figure 3.6.  The results show that mercury levels in the
digestive gland of this mussel attain equilibrium with
ambient concentrations quite rapidly; it may be significant
in this connection that mercury uptake was mainly from
particulates in the surrounding water.  By contrast,
adductor muscles exhibit much slower uptake of the metal and
may not have equilibrated completely even over the 24-week
study period.  Concentrations of mercury in the gonad may
have equilibrated prior to the large decrease seen in weeks
20 to 24, which could be related to spawning.  Such studies
on changes occurring over time in individual tissues may,
therefore, provide useful information on:

1.  relative rates of response of each tissue to changes
in ambient metal abundance,
2.  the possible dominant route of uptake of a
contaminant, and
3.  the degree to which spawning may affect the
concentrations of metals in whole soft parts of bivalves.

Finally, the presence of a highly asymmetric (non-
uniform) distribution of metals in bivalves presents
interesting possibilities concerning the study of biological
effects of pollutants.  Different tissues evidently possess
quite distinct mechanisms for the detoxification of trace
metals; thus each tissue exhibits a unique dose-response
relationship for any given metal.  Studies of the biological
effects of metals on bivalves undoubtedly call for
individual tissue analysis, so that any effect may be
correlated to the concentration of metal(s) present.

114

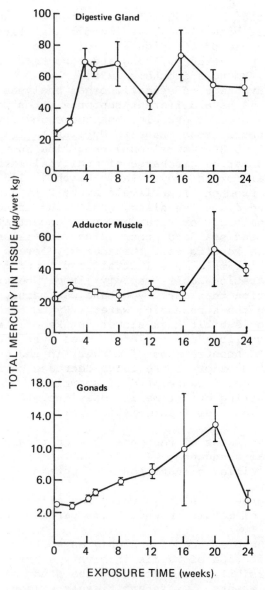

FIGURE 3.6 Concentrations of total mercury (means and standard deviations, μg/kg by wet weight) in digestive gland, adductor muscle, and gonads of *Mytilus californianus* as a function of exposure period after transplantation. (Reprinted with permission from Mar. Pollut. Bull., vol. 9, R. P. Eganhouse and D. R. Young, *In situ* uptake of mercury by the intertidal mussel *Mytilus californianus,* copyright 1978, Pergamon Press, Ltd.)

Therefore, although studies of individual tissue levels of metals are impractical for routine use, such studies would be of value in areas of special concern. Such areas would normally be those of interest in terms of the ecological or physiological effects of pollution, and there appears to be an excellent possibility of coordinated effort in these hot-spot regions.

## Shells as Indicators of Metals

The use of shells of bivalves as indicators of trace metal availability has been discussed. There are several problems concerning the use of shells.

1. No completely successful scheme for surface cleansing of bivalve shells has been found. The removal of epibiota is obviously necessary, but this is not easy to achieve without shell loss. In addition, adsorbed contaminants may also be lost in the cleaning process.

2. Analysis of shell material presents a matrix problem when atomic absorption techniques are used, especially for calcium interference in lead determination. Use of neutron activation analysis overcomes this problem, but such alternatives are expensive and not widely available.

3. The time integration of ambient metal levels may differ according to the rate of shell deposition. Little information is apparently available concerning the rate of shell deposition in bivalves and its variation with external parameters, but variation with salinity and ionic composition of ambient water might be expected.

4. Very little information is available concerning the effects of season, size, salinity, and temperature, or any other variable on the trace metal levels found in bivalve shells. However, these variables almost certainly affect metal deposition in the shell matrix, as they are known to affect metal retention by the whole soft parts of bivalves. Rucker and Valentine (1961) published data suggesting that the concentrations of several metals in the shells of Crassostrea gigas were affected by salinity.

Thus, even apart from the practical difficulties of using shells as indicators of trace metals, the lack of information on external variables diminishes the value of this technique. However, there are scientists who feel that shells could be used as indicators if enough research were devoted to investigation of the effects of the major variables. Sturesson (1976, 1978) has published details of experiments with lead and cadmium that suggest that the shells of _Mytilus edulis_ may be useful as indicators. The study of shells is therefore considered at this time to be more suited to a research project than to a monitoring program.

## Aging Methods

Recent research suggests that shell thickness in mussels may be a useful diagnostic tool in determining the age of the organism (Griffin and others 1980). Shell thickness is measured (in mg/cm$^2$) by determining both the dry weight of the shell and its area (by pressing aluminum foil over the shell, cutting around the outline, and weighing the foil). This parameter may be an improvement on simple shell length measurement, as the differences in shell shape of mussels from location to location lead to inaccuracies in comparisons of age based on length of shell.

Furthermore, Griffin and others (1980) identified a linear relationship between the (log) of the content of $^{210}$Pb and $^{210}$Po in the whole soft parts of _Mytilus edulis_ and _M. californianus_ and their shell thicknesses, measured as noted above. Consideration of these data suggests that the parameters may all have value as methods of aging mussels. However, there is some evidence of location-dependence in the data for mussels from the Pacific coast of the United States; thus the slopes of regressions relating shell thickness to either the number of lines in the shell (determined by an acetate peel technique) or to the $^{210}$Pb or $^{210}$Po contents of the whole soft parts appear to vary between stations. The most likely cause of the variation would be differences in the deposition rate of the shell between mussels at different sites. In any event, such site-dependence of the correlations may eliminate accurate aging of mussels by these techniques (for all stations) by comparing results at one station to a master regression. Clearly, further research is needed before this is used as a routine tool in monitoring, but in areas of particular interest, such methods may be useful as a means of aging mussels or other bivalves.

## Transplantation of Bivalves

Earlier in this report, bivalves were identified as the best candidates for an international monitoring survey. In some areas of the world, however, few bivalves occur naturally. The eastern Mediterranean Sea, for example, possesses only a few scattered and isolated populations of Mytilus galloprovincialis. In addition, the distribution of a bivalve is sometimes found to be inadequate for intensive studies of highly polluted hot-spot areas. In such cases, the use of transplantation techniques may be valuable. Bivalve transplantation from one area to another is not particularly difficult. Early studies employed such a technique to investigate the half-lives of radionuclides in bivalves, and several authors have since transplanted bivalves into areas of known contamination (e.g., in Western Port Bay, Australia; Poole Harbour, Great Britain; Narrangansett Bay, United States; Oslofjord, Norway). The main constraints on such efforts are the design of a suitable tray to hold the transplanted animals and the need to avoid extreme salinity or temperature shock during transplantation. In addition, it should be noted that several months or longer may be necessary after transplantation for metal concentrations present in bivalves to equilibrate to new environmental levels. Equilibration of each tissue takes a certain unique period correlating to the rate of metal exchange through the tissue (metal flux). For tissues such as the digestive gland, equilibration is rapid, whereas in adductor muscle, it is extremely slow (Pentreath 1973). If whole soft parts are to be monitored, results should be used only after a period sufficient to allow equilibration of the tissue exhibiting the slowest rate of metal exchange. If this period is not known, an organism to be transplanted into a barren area for study should also be transplanted into an area that contains native animals of the same species but where the availability of metal is different from the area of origin of the transplant. It may then be assumed that equilibration in the two new environments will proceed at approximately similar rates, and comparison of native and transplanted bivalves in the "control" area will define equilibration in the previously barren study area. Complete equilibration of transplants may take up to a year or more in some instances; thus in areas where transplantation is needed, a sufficient lead-in time after transplantation must be permitted before monitoring can be started.

## Half-Lives of Trace Metals in Bivalves

Consideration of the half-lives of trace metals in bivalves is important, as the rate of metal excretion may define, to a large extent, the length of time between successive samplings in an ongoing monitoring survey. If metal fluxes are high, sampling should be frequent to permit continual monitoring; conversely, if metals are lost slowly, less frequent sampling would adequately represent metal availability at all locations.

The amount of data published that concerns the half-life of metals in bivalves is not great and, for the most part, concerns radioactive metal isotopes. In addition, the caution recommended by Fowler and others (1975) should be kept in mind; that is, the apparent biological half-lives of metals in any given species vary according to exposure route and to length of exposure. However, general estimates are available.

Phillips (1977b) has stated that bivalves exhibit less time-integration than macroalgae, based on estimates of half-lives for several elements. Seymour (1966), using transplantation techniques, reported the loss of $^{65}Zn$ from Crassostrea gigas at an average of 0.55 percent per day, and the rate was relatively constant with season or year of study. The decay-corrected biological half-life of $^{65}Zn$ was estimated to be 255 days, corresponding to an effective half-life (not corrected for radioactive decay) of 125 days. Young and Folsom (1967) transplanted the mussel Mytilus californianus from the Columbia River to La Jolla, California, and found a biological half-life for $^{65}Zn$ of 76 $\pm$ 3.5 days. Seymour and Nelson (1973) followed the decline of $^{65}Zn$ in both M. edulis and M. californianus in situ in the Columbia River after shutdown of the Hanford reactors in February 1971. Ecological half-lives for the species were found to be 277 and 380 days, respectively. The latter values are much greater than those of Young and Folsom (1967). The discrepancy is probably a function of two parameters.

1. The data of Seymour and Nelson (1973) were actually reported as "ecological half-lives," a term introduced for use when concomitant uptake of a contaminant is occurring during estimation of its biological half-life.
2. No data are available concerning growth dilution effects in the two instances. If metals (or radionuclides) are measured in concentration terms and the differences between the two converted to a biological half-life,

apparent decreases in the contaminant may be caused by growth of the organism as well as by loss of contaminant. A more accurate method of comparison in these cases might involve use of the change in total body burden of the contaminant with time.

Wolfe (1970) also quoted results in terms of an ecological half-life, and related $^{65}Zn$ activity to total zinc in _Crassostrea gigas_ from Beaufort, North Carolina. In this case, estimates for the ecological half-life were 347 days for absolute activity of $^{65}Zn$ and 276 days for specific activity, the difference being due to increased concentration of the stable element during the period of $^{65}Zn$ excretion.

The only metal other than zinc for which sufficient data exist to permit a reliable estimate of half-life in bivalves is mercury. Reports concerning the half-life of mercury in bivalves present values ranging from 9 to 36 days for _Crassostrea virginica_ after an exposure period of 19 to 45 days using mercuric acetate (Cunningham and Tripp 1973, 1975a,b) to 481 days for methylmercury in _Tapes decussatus_ (Unlu and others 1972) and 1,000 days for methylmercury in _Mytilus galloprovincialis_ (Miettinen and others 1972). Such wide variation reflects not only differences between species and metal forms, but also the effects of exposure routes on metal kinetics in bivalves.

It is therefore difficult to reach reliable conclusions concerning the half-lives of trace metals in bivalves. Current estimates of half-lives often vary over two orders of magnitude, and this is as much a function of experimental technique as any reflection of the real situation. Accurate estimates of metal half-life are difficult to produce, because no completely uncontaminated environment exists for experimentation, and the effects of covarying parameters may be substantial. In addition, an observed half-life is valid only for the particular species/metal pair considered, as the excretion rate of any metal is unique in any bivalve species. Notwithstanding these uncertainties, it is probably valid to assume half-lives of at least several months for trace metals in bivalves in the environment. Given this degree of time-integration, it would appear reasonable to sample annually in large-scale monitoring surveys. Annual sampling was employed both in the U.S. Mussel Watch Program (Goldberg and others 1978) and in other programs such as ICES and Scandinavian surveys (Phillips 1977c, 1978a, 1979a). In addition, annual sampling for general wide-scale monitoring is compatible with constraints

placed on sampling by seasonal variation in trace metal
levels in bivalves. Monitoring of hot-spot areas could be
more frequent, especially where episodic inputs (e.g., from
industry) are suspected, although monitoring more frequently
than 2 to 4 times in any year would appear unnecessary even
in these areas.

## CONCLUSIONS ON SAMPLING AND ANALYSIS PROCEDURES

This chapter discusses theoretical and practical aspects
of the use of bivalve molluscs as indicators of trace metals
in marine and estuarine environments. An attempt has been
made to define parameters that may significantly interfere
with the ability of bivalves to accurately reflect the
ambient abundance or availability of trace metals. On the
basis of our current knowledge of these parameters and their
effects, sampling and analytical procedures have been
determined that permit the use of available manpower, money,
and expertise with the least wastage. The general
monitoring strategy outlined below is thus a compromise
between scientific desire and practical or economic
possibilities.

We envisage a three-tier strategy for monitoring trace
metals in bivalves from the oceans and coastal waters of the
world. The first tier comprises a general survey of both
offshore and nearshore waters in an attempt to define areas
of possible trace metal contamination or pollution. A lead-
in period of at least 6 months or 1 year prior to the
commencement of sampling is necessary, during which
analytical intercalibration exercises will be performed and
personnel will be given the necessary training. It is
assumed that relevant samples for analytical
intercalibration will be available throughout the lead-in
period as well as during subsequent sampling periods.

Bivalve species to be used in the general first-tier
survey would be those of the genera _Mytilus_, _Crassostrea_,
and _Saccostrea_. Species of particular interest, owing to
their wide distribution, are _Mytilus_ _edulis_, _M._
_galloprovincialis_ and _M._ _californianus_; _Crassostrea_ _gigas_,
_C._ _virginica_ and possibly _C._ _angulata_; _Saccostrea_ _cuccullata_
and _S._ _glomerata_. Areas that do not support wild or
cultivated populations of any of these species (or closely
related species) should have transplants deployed in study
sites at least 9 months to 1 year prior to the commencement
of sampling for monitoring purposes. The number of species

used worldwide should be kept at a minimum to allow maximum cross-comparison of data.

Sampling for the general first-tier monitoring survey should be performed in winter in temperate zones, and subsequent to the high runoff period in tropical zones, unless local knowledge of the sexual cycle of the selected bivalve dictates otherwise. Twenty-five individuals should be taken from each study location as an absolute minimum, with shell lengths (longest dimension) within the ranges of 20 to 30 mm for mussels and 25 to 35 mm for oysters, the averages for each site being as close to 25 and 30 mm, respectively, as possible. The individuals should be collected from the mid-tidal zone, halfway between marks for spring low and high waters. Each location should be photographed if possible, so that the pattern of return visits can be identical in future surveys. All locations in any local survey, or on any one coast, should be sampled in as short a time as practical. Sites with unusual hydrological conditions, such as those in heated outfalls of industry, should be avoided unless the specific intention is to monitor that industry. All collection sites should be studied over a reasonable period of time to allow for reliable estimates of parameters like salinity and water temperature.

Immediately after collection, bivalve samples should be cleaned of epibiota or adhering sediment and placed in 20 liters of water from the collection site, the water having first been allowed to settle, if necessary, to minimize its content of particulates. Samples should be maintained in this water (adequately aerated and protected by reasonable measures against contamination) for at least 24 hours and at most 48 hours. The depuration procedure may be performed in the field, but laboratory conditions are preferable. Subsequent procedures require laboratory facilities adequate to avoid contamination of samples.

After depuration is complete, individual bivalves should be wedged open with perspex wedges to drain. Any sediment adhering to the mantle or gill edges should be removed by jets of the water used during depuration. The adductor muscle should not be torn during wedging, nor should tissues be disturbed more than necessary. Draining should be permitted for 5 minutes; wedges should then be removed and samples deep frozen (-10 to -20°C) in clearly labelled polyethylene bags for storage. Any samples dispatched from a collection site to a laboratory for analysis must be kept frozen during dispatch.

After storage, samples should be thawed and individually
shucked. Stainless steel implements are not considered
significant sources of contamination during shucking.
Shells should be kept for possible age determination, and
may be refrozen for storage. All liquor in each individual
should be treated as part of the sample for analysis. After
bulking of samples (25 individuals in each bulk sample), wet
weights should be recorded to at least two significant
figures. If homogenization is desired, an ultrasonic cutter
should be employed, with precautions taken to avoid
contamination. An aliquot may be taken after complete
homogenization for the determination of a dry-weight/wet-
weight ratio. Alternatively, the entire sample may be dried
and the dry weight determined. Finally, the sample should
be subjected to the selected digestion and analytical
procedures, the necessary standards and blanks being run
concurrently. Analytical results should be quoted on both
wet-weight and dry-weight bases.

The above procedural notes concern the first-tier
general monitoring survey, covering large areas. All parts
of the sampling and analytical processes have been kept as
simple and rapid as possible, and it is recommended that all
possible capability be put into maximizing the sample (site)
number studied. Most of the work load in the general survey
occurs during sampling and early preparative stages, with
relatively little effort needed in analysis. This type of
survey is ideal as a primary investigation, aimed at general
delineation of areas of high metal availability. Second-
tier and third-tier investigations, described below, are
envisaged as methods of identifying point sources of trace
metals and investigating their ecological and physiological
significance.

A second-tier survey would be performed in areas
identified from the primary general survey as those of high
metal availability. Sampling and analysis procedures would
be essentially unchanged from those of the primary survey
(although fewer laboratories would be involved in
coordination of any second-tier survey study), but the
number of study sites per unit length of coastline would be
increased markedly, with a greater emphasis on selection of
sites to monitor individual outfalls or estuaries. In this
way, a clearer definition of a hot-spot area or point source
would emerge. It should be noted that a second-tier survey
would cover a smaller area of coastline than the primary
survey. Constraints on sampling because of nonsynchronous
spawning of bivalve populations would therefore be much
reduced, and second-tier sampling could probably be

performed at times other than those recommended for primary surveys. To be successful, second-tier surveys should employ the same bivalve species throughout the study area; transplantation may be needed for some sites.

Third-tier investigations would be initiated on areas identified by second-tier studies as hot-spots or point sources of metals. At this stage it is assumed that monitoring of metal levels would proceed in cooperation with studies on the ecological and physiological effects of the contamination. There may be need for the analysis of individual tissues of bivalves, of individual bivalves (in order to study size/concentration relationships), of bivalves from different vertical positions on the shoreline (to study metal stratification in the water column), and so on. The exact studies needed would depend on the nature of the point source, the hydrology and oceanography of the area, and on other factors, such as whether the metal input is continuous or episodic in nature. Conceivably, the use of other types of indicator organisms, such as macroalgae or crustaceans, may be considered worthwhile at this stage.

To conclude, bivalve monitoring can be a powerful tool for the identification and study of areas polluted by trace metals. The success of large-scale investigations such as those in the United States, parts of Europe, Scandinavia, and the Mediterranean, shows that international or worldwide monitoring of bivalves for trace metals could become a reality if the necessary capital investment and international cooperation are forthcoming. Such programs could yield invaluable information on the health of coastal and oceanic waters and could serve as a baseline for comparison with future levels.

## APPENDIX 3-A

### SUBJECTS FOR FUTURE RESEARCH

Certain aspects of the problems considered by the trace metals panel in connection with worldwide monitoring of trace metals in bivalves were deemed particularly worthy of future research, possibly in cooperation with agencies currently involved in bivalve monitoring. These subjects are outlined below, the order corresponding to that in the preceding report.

## Arsenic Speciation

If arsenic is to be included in a monitoring program using bivalves as indicator organisms, it is suggested that the different arsenic species be considered separately, as the inorganic arsenic species (arsenate and arsenite) and the organic arsenic species (monomethyl arsenic acid and dimethyl arsonic acid) have different effects on different living organisms. The development of techniques to quantify separately the different species of arsenic present in bivalves is thus of value. The speciation of other metals in aquatic biota, e.g. lead and tin, is also of interest; however, research into these other metals is currently not as advanced as that for arsenic, and development of our understanding is thus more of a long-term proposition for lead and other metals.

## Species Taxonomy

Considerable doubt exists concerning speciation in oysters of the genera Crassostrea and Saccostrea, and some doubt even remains concerning speciation in the genus Mytilus. The currently accepted situation is that common oysters may be split into tuberculated or denticulated species, which are ascribed to the genus Saccostrea, and those with no such characters, ascribed to the group Crassostrea. However, some confusion exists concerning nomenclature, partly due to the naming of Saccostrea-like oysters in Australia and New Zealand as Crassostrea commercialis (Saccostrea cuccullata) or Crassostrea glomerata (Saccostrea glomerata).

Once such a division is accepted, the problem of speciation within each genus arises. Thus the differences between, for example, Mytilus edulis, M. galloprovincialis and M. californianus are frequently indistinct, as are those between Crassostrea gigas and C. angulata; species of Saccostrea are also difficult to separate taxonomically. This subject is most important with respect to trace metal monitoring, as species differences in trace metal accumulation may be severe and will confuse interpretation of monitoring results if the bivalve samples are incorrectly identified taxonomically.

We suggest three possible approaches to the clarification of bivalve taxonomy as it relates to the use of bivalves for trace metal monitoring.

1. Standard taxonomic procedures may be used, ranging from morphological comparisons, to cross-fertilization studies, to comparison of biochemical or genetic traits.  It is important to note here that differences between populations of two supposedly different species must be demonstrably greater than those between individual populations of the same species.

2. As trace metal accumulation is species-specific, the concentrations of metals in bivalves may be used as taxonomic characters.  Thus we suggest studies of bivalves in areas of overlapping distribution of two or more species, to elucidate differences in trace metal contents.  If two species coexist in slightly different habitats (e.g., Mytilus edulis and M. californianus), they should be transplanted into the same habitat prior to study, allowing a reasonable period for reequilibration of trace metal levels.

3. In addition to the field studies described in two above, bivalves of supposedly different species could be studied for metal uptake in the laboratory.  If uptake rates from water or food--or both--differ, it may be concluded that the bivalves should be considered to be of different species at least from the standpoint of monitoring for trace metals.


## Studies on Saccostrea Species

Our knowledge concerning the effects of variables on the accumulation of trace metals by species of Saccostrea is much less advanced than that for species of either Mytilus or Crassostrea.  Thus the published literature to date includes only those papers on Saccostrea cuccullata from the east Australian coast (Mackay and others 1975) and on S. glomerata from New Zealand (Nielsen and Nathan 1975).  Phillips (1979b) has recently concluded studies aimed at testing the indicator ability of S. glomerata in Hong Kong waters.  However, further studies are needed to investigate the effects of variables such as season, tidal position, and salinity or temperature on the uptake of metals by oysters of this genus.


## Sexual Differences in Trace Metal Content

The current knowledge concerning sex-based differences in trace metal contents of bivalves is inadequate.  Further

studies are needed to investigate the possible effects of
sexual differences on bivalves to be used in a worldwide
monitoring program.  Species for study should be Mytilus
edulis, M. galloprovincialis, M. californianus, Crassostrea
gigas, C. virginica, C. angulata, Saccostrea cucullata, and
S. glomerata, as these eight species are considered to be
those most likely to be used in international bivalve
monitoring.  It should be noted that if sex-based
differences in trace metal contents are found to be severe
in any of these species, an extra restraint on sampling will
be necessary in order to eliminate such an effect.  In this
connection, it was noted previously that differences in the
trace metal levels found in different sexes may well be
seasonally dependent.  Thus studies of such differences
should be performed on bivalves at the time of sampling for
the proposed monitoring survey (winter in temperate
climates, and subsequent to the high run off period in
tropical climates).  As oysters and (particularly) mussels
may both contain ripe gonads at these sampling times, the
possibility of detecting sex-based differences in trace
metal levels appears high.  For first-tier and second-tier
surveys we would be concerned only with significant effects
on the levels of metals in whole soft parts of bivalves.
However, third-tier surveys, in which different tissues
would possibly be analyzed separately, should consider the
sex of bivalves as an important parameter, and bivalve
samples should be sexed as a matter of course in third-tier
surveys.  It would also be of value to analyze gametes of
bivalves in such a research program, as this would not only
be relevant to the understanding of sex-based differences
but would also be an important contribution to our knowledge
of the seasonal fluctuation of trace metals in different
species of bivalves.


Inherent Variability

It was noted earlier that current data concerning the
magnitude of so-called "inherent variability" in trace metal
concentrations exhibited by bivalves are inadequate, in that
data are available for only a few species/metal pairs.
Further research is therefore needed in this area, for each
of the eight species suggested as candidates for a global
monitoring survey (see the previous section on sexual
differences).  This research entails the analysis of perhaps
50 similarly sized individuals of a given species, collected
from the same location at a season other than that of

spawning.  These individuals should at least be sexed;
ideally, 50 individuals of each sex should be analyzed.
Should inherent variability be of great magnitude in any one
species or species/metal pair, the suggested minimum sample
size (25 individuals in a bulk homogenate) may have to be
increased for that species.


## Effects of Depuration

The panel members agreed that on the basis of present
knowledge, all samples should undergo depuration to
eliminate gut contents prior to analysis.  However, it is
recognized that this procedure is often difficult to set up
in the field, and presents logistical problems.  We
therefore suggest that studies designed to investigate the
loss of trace metals during depuration are initiated, using
species considered as the major candidates for worldwide
monitoring.  If these studies show that the loss of all
metals of interest during depuration is less than 5 to 10
percent, researchers may dispense with depuration.  It is
clearly not sufficient to show that the loss of any one
metal is minimal, however.


## Volatility of Metals

Members of the panel were concerned that some metals
might be partially lost during sample preparation by
volatilization during the drying procedure.  If the sample
for drying is an aliquot of a bulk homogenate, used solely
to produce a dry weight, this problem does not arise, as
analysis would be performed using an undried portion of the
sample.  However, in third-tier surveys at least, we
envisage the analysis of individuals, and these must be
dried prior to analysis.  Thus research into the possible
effects of drying samples at 70 to 105°C would appear
necessary, at least for third-tier surveys.  We suggest that
such research could be performed during the intercalibration
lead-in phase, prior to the commencement of studies on
monitoring samples.  Laboratories with recognized expertise
should undertake such studies, using bivalve species to be
sampled in the monitoring program itself.  The effect of
sample lipid on the analysis of silver might also be
investigated at this time.

Shells as Indicators and Methods of Aging

The panel recognized that bivalve shells presented potential both as indicators of some metals (notably lead, cadmium, and some of the transuranics) and as a means of aging bivalves. However, the practical problems associated with the analysis of shell material limits their value, at least as accurate indicators, as does the current lack of knowledge concerning the effects of external variables on metal levels in bivalve shells. Further research in this area is obviously needed, preferably in areas with well-defined pollution gradients.

The panel felt that research on aging mussels by thickness or $^{210}$Pb and $^{210}$Po contents is in its infancy at present, and it would be difficult to commit a major international program to additional expense on the basis of present data. In particular, site-dependence of the parameters used in these aging methods appeared probable from the data of Griffin and others (1980), and this would severely limit the usefulness of such a method, at least in general surveys. However, aging techniques may be a useful aid to study in hot-spot areas. In this connection, several researchers have noted unusually thin shells in mussels from polluted locations or extremely hyposaline locations. Some members thus felt that shell thickness would be found to be dependent on variables other than age, and therefore the age-thickness correlation would vary with location. Tests of such location-dependent variation in the rate of shell synthesis by bivalves should be relatively simple to design and carry out.

## REFERENCES

Ahmed, M. (1975) Speciation in living oysters. Adv. Mar. Biol. 13:357-397.

Alexander, G.V. and D.R. Young (1976) Trace metals in southern California mussels. Mar. Pollut. Bull. 7(1):7-9.

Boyden, C.R. (1974) Trace element content and body size in molluscs. Nature 251:311-314.

Boyden, C.R. (1977) Effect of size upon metal content of shellfish. J. Mar. Biol. Assoc. 57:675-714.

Bryan, G.W. (1973) The occurrence of seasonal variation of trace metals in the scallops Pecten maximus (L.) and Chlamys opercularis (L.). J. Mar. Biol. Assoc. 53:145-166.

129

Cunningham, P.A. and M.R. Tripp (1973) Accumulation and
    depuration of mercury in the American oyster Crassostrea
    virginica. Mar. Biol. 20:14-19.
Cunningham, P.A. and M.R. Tripp (1975a) Factors affecting
    the accumulation and removal of mercury from tissues of
    the American oyster Crassostrea virginica. Mar. Biol.
    31:311-320.
Cunningham, P.A. and M.R. Tripp (1975b) Accumulation, tissue
    distribution and elimination of $^{203}HgCl_2$ and $CH_3$ $^{203}Hg$
    Cl in the tissues of the American oyster Crassostrea
    virginica. Mar. Biol. 31:321-334.
De Wolf, P. (1975) Mercury content of mussels from west
    European coasts. Mar. Pollut. Bull. 6(4):61-63.
Eganhouse, R.P. and D.R. Young (1978) In situ uptake of
    mercury by the intertidal mussel Mytilus californianus.
    Mar. Pollut. Bull. 9(8):214-217.
Food and Agriculture Organization (1974) Technical Report
    No. 158. Rome: Food and Agriculture Organization.
Fowler, S.W. and B. Oregioni (1976) Trace metals in mussels
    from the north-west Mediterranean. Mar. Pollut. Bull.
    7(2):26-29.
Fowler, S.W., J. La Rosa, M. Heyraud, and W.C. Renfro (1975)
    Effect of different radiotracer labelling techniques on
    radionuclide excretion from marine organisms. Mar. Biol.
    30:297-304.
Frazier, J.M. (1975) The dynamics of metals in the American
    oyster, Crassostrea virginia: I. Seasonal effects.
    Chesapeake Sci. 16:162-171.
Frazier, J.M. (1976) The dynamics of metals in the American
    oyster, Crassostrea virginica II: Environmental effects.
    Chesapeake Sci. 17:188-197.
Goldberg, E.D., V.T. Bowen, J.W. Farrington, G. Harvey, J.H.
    Martin, P.L. Parker, R. W. Risebrough, W. Robertson, E.
    Schneider, and E. Gamble (1978) The mussel watch.
    Environ. Conserv. 5(2):101-125.
Gordon, M., G.A. Knauer, and J.H. Martin (1978) Intertidal
    Study of the Southern California Bight. Manuscripts
    originating from the Moss Landing Marine Laboratory,
    California.
Griffin, J.J., M. Koide, V. Hodge, and E.D. Goldberg (1980)
    Determining the ages of mussels by chemical and physical
    methods. In Isotope Marine Chemistry, 70th Anniversary
    volume for Y. Miyake, Meteorological Institute. Tokyo
    (in press).
Harrison, F.L. (1978) Effect of the physicochemical form of
    trace metals on their accumulation by bivalve molluscs.

Preprint UCRL-81134, submitted to the American Chemical
    Society Annual Meeting, Miami, Florida, September 1978.
International Decade of Ocean Exploration (1973) Workshop
    Report on Baseline Studies of Pollutants in the Marine
    Environment and Research Recommendation. IDOE Baseline
    Conference, New York, May 24-26, 1972.
Johnels, A.G., T. Westermark, W. Berg, P.I. Persson, and B.
    Sjostrand (1967) Pike (Esox lucius L.)and some other
    aquatic organisms in Sweden as indicators of mercury
    contamination in the environment. Oikos 18:323-333.
Mackay, N.J., R.J. Williams, J.L. Kacprzac, M.N. Kazacos,
    A.J. Collins, and E.H. Auty (1975) Heavy metals in
    cultivated oysters (Crassostrea commercialis =
    Saccostrea cucullata) from the estuaries of New South
    Wales. Aust. J. Mar. Freshwater Res. 26:31-46.
Majori, L., G. Nedoclan, G.B. Modonutti, and F. Davis (1978)
    Study of the seasonal variations of some trace elements
    in the tissues of Mytilus galloprovincialis taken in the
    Gulf of Trieste. Rev. Int. Oceanogr. Med. 49:37-40.
Martin, J.H. and D. Phelps (1979) Bioaccumulation of Heavy
    Metals by Littoral and Pelagic Marine Organisms. EPA
    Final Report No. 600/3-79/038. Washington, D.C.:   U.S.
    Environmental Protection Agency.
Miettinen, J.K., M. Heyraud, and S. Keckes (1972) Mercury as
    a hydrospheric pollutant: II. Biological half-time of
    methyl mercury in four Mediterranean species:  A fish, a
    crab and two molluscs. Pages 295-298, Marine Pollution
    and Sea Life, edited by M. Ruivo. Surrey, England:
    Fishing News (Books) Ltd.
Nielsen, S.A. (1974) Vertical concentration gradients of
    heavy metals in cultured mussels. N. Zeal. J. Mar.
    Freshwater Res. 8(4):631-636.
Nielsen, S.A. and A. Nathan (1975) Heavy metal levels in New
    Zealand molluscs. N. Zeal. J. Mar. Freshwater Res.
    9(4):467-481.
Ouellette, T. (1978) Seasonal Variation of Trace Metals and
    the Major Inorganic Ions in the Mussel Mytilus
    californianus. Master's thesis, California State
    University, Hayward.
Pentreath, R.J. (1973) The accumulation from water of $^{65}$Zn,
    $^{54}$Mn, $^{58}$Co and $^{59}$Fe by the mussel, Mytilus edulis. J.
    Mar. Biol. Assoc. 53:127-143.
Phillips, D.J.H. (1976a) The common mussel Mytilus edulis as
    an indicator of pollution by zinc, cadmium, lead and
    copper: I. Effects of environmental variables on uptake
    of metals. Mar. Biol. 38:59-69.

Phillips, D.J.H. (1976b) The common mussel Mytilus edulis as
    an indicator of pollution by zinc, cadmium, lead and
    copper: II. Relationship of metals in the mussel to
    those discharged by industry. Mar. Biol. 38:71-80.
Phillips, D.J.H. (1977a) Effects of salinity on the net
    uptake of zinc by the common mussel Mytilus edulis. Mar.
    Biol. 41:79-88.
Phillips, D.J.H. (1977b) The use of biological indicator
    organisms to monitor trace metal pollution in marine and
    estuarine environments: A review. Environ. Pollut.
    13:281-317.
Phillips, D.J.H. (1977c) The common mussel Mytilus edulis as
    an indicator of trace metals in Scandinavian waters: I.
    Zinc and cadmium. Mar. Biol. 43:283-291.
Phillips, D.J.H. (1978a) The common mussel Mytilus edulis as
    an indicator of trace metals in Scandinavian waters: II.
    Lead, iron and manganese. Mar. Biol. 46:147-156.
Phillips, D.J.H. (1978b) Use of biological indicator
    organisms to quantitate organochlorine pollutants in
    aquatic environments: A review. Environ. Pollut. 16:167-
    229.
Phillips, D.J.H. (1979a) Trace metals in the common mussel,
    Mytilus edulis (L), and in the alga Fucus vesiculosis
    (L.) from the region of the sound (Oresund). Environ.
    Pollut. (in press).
Phillips, D.J.H. (1979b) The rock oyster Saccostrea
    glomerata as an indicator of trace metals in Hong Kong.
    Marine Biol. 53:353-360.
Romeril, M. Paper submitted to Estuarine and Coastal Marine
    Science (in press).
Rucker, J.B. and J.W. Valentine (1961) Salinity response of
    trace element concentration in Crassostrea virginica.
    Nature 190:1099-1100.
Seymour, A.H. (1966) Accumulation and loss of zinc-65 by
    oysters in a natural environment. In Disposal of
    Radioactive Wastes into Seas, Oceans and Surface Waters.
    IAEA Report SM-72/38. Vienna: International Atomic
    Energy Agency.
Seymour, A.H. and V.A. Nelson (1973) Decline of $^{65}$Zn in
    marine mussels following the shutdown of Hanford
    reactors. In Radioactive Contamination of the Marine
    Environment. IAEA Report SM-158/16. Vienna:
    International Atomic Energy Agency.
Stenzel, H.B. (1971) Oysters. In Treatise on Invertebrate
    Paleontology, N 953-N. Part N, Vol. 3, Mollusca 6,
    edited by K.C. Moore. Boulder, Colo: Geological Society
    of America Inc. and the University of Kansas.

132

Stephenson, M.D., J.H. Martin, and M. Martin (1978) State
     Mussel Watch, 1978 Annual Report. Trace Metal
     Concentrations in the California Mussel at Areas of
     Special Biological Significance. Manuscript originating
     from the Moss Landing Marine Laboratory, California.
Sturesson, U. (1976) Lead enrichment in shells of *Mytilus
     edulis*. Ambio 5:253-256.
Sturesson, U. (1978) Cadmium enrichment in shells of *Mytilus
     edulis*. Ambio 7:122-125.
Topping, G. and A.V. Holden (1978) ICES Report on
     Intercalibration Exercises. Cooperative Research Report
     No. 80, March. Charlottenlund, Denmark: International
     Council for the Exploration of the Sea.
Unlu, M.Y., M. Heyraud, and S. Keckes (1972) Mercury as a
     hydrospheric pollutant: I. Accumulation and excretion of
     $^{203}HgCl_2$ in T*apes decussatus* L. Pages 292-295, Marine
     Pollution and Sea Life, edited by M. Ruivo. Surrey,
     England: Fishing News (Books) Ltd.
Watling, H.R. (1978) Selected Molluscs as Monitors of Metal
     Pollution in Coastal Marine Environments. Ph.D. thesis,
     University of Cape Town, South Africa.
Watling, H.R. and R.J. Watling (1976) Trace metals in
     *Choromytilus* *meridionalis*. Mar. Pollut. Bull.
     7(5):91-94.
Wolfe, D.A. (1970) Levels of stable Zn and $^{65}$Zn in
     *Crassostrea* *virginica* from North Carolina. J. Fish. Res.
     Bd. Canada 27:47-57.
Wong, P.T.S., Y.K. Chau, and P.L. Luxon (1975) Methylation
     of lead in the environment. Nature 253:263-264.
Young, D.R. and T.R. Folsom (1967) Loss of Zn-65 from the
     California sea-mussel *Mytilus californianus*. Biol. Bull.
     133:438-447.
Young, D.R., G.V. Alexander, and D. McDermott-Ehrlich (1978)
     Vessel related contamination of southern California
     harbors by copper and other trace metals. Mar. Pollut.
     Bull. (in press).

# 4

# HALOGENATED HYDROCARBONS

ALAN V. HOLDEN (Chairman), Department of Agriculture and
    Fisheries for Scotland, Perthshire, Scotland
MARGUERITA BARROS, Labratorio de Farmacologia, Oeiras,
    Portugal
PHILIP A. BUTLER, Environmental Protection Agency, Gulf
    Breeze, Florida
EGBERT G. DUURSMA, Royal Netherlands Academy of Arts and
    Sciences, Yerseke, Netherlands
GEORGE HARVEY, U.S. Department of Commerce, Miami, Florida
MICHEL MARCHAND-STAVRE, Centre Oceanologique de Bretange,
    Cedex, France
G. BALUJA MARCOS, Institute of Organic Chemistry, Madrid,
    Spain
ILKAY SALIHOGLU, Middle East Technical University, Icel,
    Turkey

## INTRODUCTION

Halogenated hydrocarbons, and particularly DDT, DDE, and
PCBs, are transported to a significant extent through the
atmosphere and have been detected in all regions of the
world.  The substances, which have been produced only by
man, are now as widely distributed as natural substances,
although, in the case particularly of the PCBs, their use is
confined to limited areas of the world.  Natural processes
of atmospheric transport lead inevitably to the
redistribution of halogenated hydrocarbons to areas where no
use exists or can be anticipated, and it is to be expected
that in these areas the background level will increase
slowly, although not necessarily to concentrations at which
biological effects could occur.  Nevertheless, it may be
considered prudent to monitor concentrations in such areas
to assess the extent of contamination and to measure the

133

trend in concentrations of substances such as PCBs, DDT, and
DDE over a long period.

The panel discussed both the question of alternatives to
bivalves as material for assessing pollution in coastal
waters and strategies required for determination of any
organohalogen contamination in the samples selected. As the
organohalogen group includes several hundred compounds of
various types and uses, and of widely differing chemical
properties, the panel decided that only a limited number of
more persistent compounds with widespread occurrence should
be measured in any monitoring program. The specific
compounds will be determined by use patterns in the
respective areas, although in all cases PCBs should be
analyzed.


## BIVALVES AS MONITORING SPECIES

The panel discussed the relative merits of mussels or
oysters as indicators of coastal pollution as compared with
alternatives such as macroalgae, fish, or water. Macroalgae
may accumulate material on the outer surface, including both
absorbent particulate matter and petroleum HCs, which may
itself contain organohalogenated compounds and thus give
excessively high concentrations in the analytical
determination. Algae may have relatively low lipid content,
so that the concentrations of the lipid-soluble
organohalogens within samples could be low. Macroalgae are
therefore not particularly effective as indicators or
integrators of ambient pollution.

Fish may be good accumulators of organohalogens if they
have a high lipid content, but they are mobile, often
difficult to sample, and may not reflect local
contamination. Most species of fish are not good
experimental animals and the factors controlling pollutant
uptake and excretion, and the effects of the pollutants, are
difficult to assess.

In contrast, bivalves such as mussels and oysters are
easy to sample, occur very widely in most, if not all,
coastal areas of the world, and have been studied in some
detail with many pollutants. The variability among
individuals may present some problems in both sampling and
interpreting the analytical data but this situation applies
to all biological specimens and may, in the case of
measuring many organohalogens, be allowed for, to some
extent, by expressing concentrations on a lipid basis.

Bivalves concentrate the contaminant from the surrounding environment (in food particles, sediments, and water). Unless the bivalves are themselves of interest as human food, as components of the food chain of other species, or as indicators of biological effects of the pollutants, one might determine concentrations in the water itself, by the use of resin columns or polyurethane foam through which a standard volume of water is passed. The absorbent material would be uniform and thus more reproducible, eliminating the variability of biological specimens. However, the concentrations determined after extraction would be lower than those occurring in bivalves unless large volumes of water are processed. Particulate matter will be trapped on the foam plug material, but may pass through resin columns unless a fine filter is placed at the top of the column. The particulate matter can be analyzed separately or together with the water.

The panel felt that information from bivalves was probably more valuable than that obtained from examining the water itself, but the relatively low cost of analyzing a resin or foam sample compared to the cost of analyzing individual mussels, or replicates of a homogenate, justified the acquisition of the additional information from the water. It should be remembered that the absorbent column would provide information only on an instantaneous water sample unless a special collecting apparatus were used for the water sample, while the bivalves will provide an integrated value over a period of time.

The panel recognized that bivalves vary in size between sites, but information available suggested that for organochlorines the variation in concentration with size on a lipid basis within a single population was small. Smaller mussels may contain higher concentrations of organochlorines but also higher concentrations of lipid. Removal of sand and free water from the interior of the shell is considered to be important to reduce the error in subsequent weighing of the tissue in the wet or dry form. Depuration would not appear to cause any significant loss of organochlorines by excretion of the gut contents, but in any event it is not necessary for the analysis of these compounds. Dissection and analyses of individual organics is not practicable for routine monitoring. The entire soft parts (without loss of hemolymph) should be taken for analysis.

The number of individuals needed to represent adequately a single population depends in part on the accuracy required. The variation among individuals may be sufficiently expressed by analysis of perhaps 25 specimens.

Duplicate analyses will allow the assessment of the
analytical precision. With a skewed distribution among the
individual values, the homogenate method may not provide an
adequate estimate of the median value, although the error
could be acceptable and the cost of analyses thus reduced.
This subject may need more research if greater precision is
required.

Bivalves should be sampled 2 months before spawning, if
possible, to ensure that an unexpectedly early spawning does
not cause the loss of organochlorines before sampling takes
place. In some areas, mussels spawn over an extended
period, and individual spawning times must be identified
beforehand.

In seriously contaminated areas that are to be studied
in greater detail, sampling at intervals during the year may
be required. For global monitoring, samples should be taken
annually or perhaps even less frequently. The biological
half-life of organochlorines varies with the species and
compound, but is usually on the order of weeks.

The panel had no information on the spatial extent that
a sample of mussels might represent. However, it was
thought that in unpolluted waters a sample would be typical
of a large area, whereas in polluted waters the
representation could be of very small areas, possibly
requiring a greater spatial frequency of sampling.

The use of planted mussels of known age and history was
thought to be an advantage in some respects, but care must
be exercised to select environments where physical
conditions are reasonably similar.

## CHEMICAL ANALYSIS

The basic technique used for the analysis of
organohalogen compounds is gas chromotography using an
electron-capture detector which is particularly sensitive to
halogenated compounds but can also respond to higher
concentrations of many other compounds. Recent developments
of the basic technique that provide greater confidence in
identification of the compounds include the use of capillary
columns in the chromatographs and of mass spectrometers
coupled to the chromatographs. Detection and measurement of
the lowest concentrations of organohalogen compounds found
in the environment require extreme care in performing the
analysis, the use of very pure solvents and absorbents, and
very clean working conditions. The analyst needs a high
level of training before confidence can be placed on the

results, and regular intercalibration programs are essential in securing the necessary accuracy and precision, and in ensuring that these are maintained.

Many of the difficulties of providing the necessary analytical services could be avoided if the samples taken for analysis could be sent to a limited number of well-equipped laboratories with highly trained chemists throughout the world. These laboratories could participate in regular intercalibration programs to ensure that their analytical capabilities are comparable, and to provide a basis for comparisions of their results. The procedure would require a reliable method of transporting samples over long distances without deterioration.

Guidelines are needed to enable laboratories to establish the analytical services needed to monitor organohalogen residues in the environment and to provide data of an accuracy acceptable in international programs.

## COMPOUNDS TO BE ANALYZED

Although several hundred halogenated organics are known and are used in agriculture and industry, relatively few are regarded as very persistent and ubiquitous. These compounds must be evaluated in the analytical programs of as many countries as possible. The following list is suggested as a first objective: (1) PCBs; (2) p,p DDT, p,p TDE, p,p DDE; (3) o,p DDT; (4) dieldrin, aldrin; (5) hexachlorobenzene; (6) $\alpha, \beta$, and $\gamma$-hexachlorocyclohexane; and (7) heptachlor, heptachlor epoxide.

Some countries do not use several of these chemicals, but they should nevertheless attempt to detect and report them.

Other compounds including toxaphene, chlordane, mirex, endrin, endosulfan (thiodan), and Kepone® are used widely in certain countries only. All can be detected by the same technique, but in a mixture considerable analytical differences may arise (e.g., with PCBs and toxaphene). It is essential that each country be aware of the organochlorine chemicals it uses, so that the residues arising from contamination of the environment can be identified and determined.

The compounds mentioned so far are all highly fat soluble, relatively insoluble in water, and are concentrated in the lipids of living organisms. Other widely used chlorinated compounds that are more soluble in water but can still be concentrated in lipids include certain fungicides, and herbicides such as 2,4-D and 2,4,5-T derivatives and pentachlorophenol. These compounds would in many cases not be determined by the basic technique required for the other group of compounds, and would necessitate a modification of the methods. It is not considered necessary to establish international monitoring programs to seek these compounds. Halogenated solvents, fire retardants, and refrigerants (such as chloroform, trichloroethylene, chlorobrommethane), which are discharged to the environment in large quantities but are very volatile and do not generally give rise to measurable concentrations in biological organisms, are not considered appropriate targets for international programs either. The technique for determining such compounds in environmental materials is (apart from the gas chromatography) largely different from that used for the pesticides and PCBs.

## BASIC PRECAUTIONS

All chemicals used in the analytical procedure must be of high purity. Solvents such as n-hexane and acetone must be pre-tested by concentrating 100 ml to 1 ml by evaporation in a stream of pure nitrogen and injecting an aliquot of the concentrate into the chromatograph. The chromatogram produced should contain no part that will interfere in the analyses of samples at the highest sensitivity. Absorbents such as Florisil, alumina, and silica must be pre-washed with solvents, such as hexane or acetone, and then dried; the blanks carried through the analytical procedure used for samples should reveal any contaminant peaks.

Glassware and other equipment used must be cleaned with strong acid, alkali, or proprietary cleaning fluids, and washed thoroughly with water and pure acetone before use. Plastic material, especially polyvinyl chloride, must never be used, nor allowed to come into contact with glassware, solvents, or other chemicals. Teflon preparation is suitable in sampling equipment or analytical operations, but otherwise metal (preferably stainless steel) equipment is required. For further information see FAO/UNEP (1976) and Goldberg (1976).

139

## ANALYTICAL PROCEDURE

Many detailed accounts of the procedure for analyzing environmental materials to identify organochlorine residues have been published, but the basic principles are:

1. Extraction of weighed sample (often ground to powder with anhydrous sodium sulfate) in a Soxhlet or similar extractor using pure petroleum ether, hexane, or hexane/acetone.
2. Clean-up of aliquot of extract from Step 1 using either solvent partition (hexane-acetonotrile or hexane-dimethyl formamide), or more commonly an absorbent column of Florisil or alumina. This stage removes lipids and pigments, but can also remove many environmental contaminants other than the organochlorines listed for the monitoring programs.
3. Pre-gas-liquid chromatography (GLC) separation of the mixture of residues on Florisil or silica columns, to provide some evidence of identity and simplify subsequent GLC analysis.
4. Injection of small (5-ml) aliquots of the eluates from Step 3 on GLC columns, which separate the contaminant residues before their identification by the electron capture detector. The signals produced are linearly proportional to the concentrations of the residues (over a wide range) and can be compared with the response from standard solutions.
5. Confirmation of identity, which can be provided by a combination of procedures. The eluates from Step 3 give some indication, but the positions of the peaks on the chromatogram from Step 4 are usually specific for individual compounds. Where compounds elute at the same or similar positions, a GLC column with a different stationary phase may provide the necessary separation. Chemical treatment before reanalysis by GLC can provide further confirmation in changing the positions of some of the peaks on the chromatogram. Peak identity is more easily confirmed with the aid of a mass spectrometer, except in the case of certain isomers that may have similar fragmentation patterns.

An alternative GLC technique, using capillary coated columns and smaller injection volumes, can separate all but a few of the compounds to be found in environmental samples. The columns are more expensive and shorter-lived, and the technique is used by very few laboratories at the present time.

The presence of PCBs or toxaphene, both of which are mixtures of a large number of individual compounds, provides chromatograms with a series of closely spaced peaks interfering with many pesticides. The silica column separation of Step 3 separates many residues from the PCBs, but DDE and the PCB group elute together. Chlordane isomers are also separated from PCBs.

Further information can be found in Holden (1973), Wells and Johnstone (1977), FAO/UNEP (1976), and AOAC (1975).

## Analysis of Water

Several techniques have been developed for the extraction of organochlorines from water using resins or polyurethane foam. Volumes of 50 or 100 liters can be passed through small columns of absorbent, which are subsequently extracted with acetone and the residues partitioned into hexane for analysis by the procedure used for biological materials. The following references should be consulted: Musty and Nickless (1974), Uthe and others (1974), Dawson and Riley (1977), Wells and Johnstone (1978), and Harvey and Steinhauer (1976).

## Lipid Concentration

Organochlorine concentrations are usually expressed in terms of the unit wet weight of tissue. Regulatory limits for contaminant levels in food are given on the same basis, and thus the information from the analysis of mussels can be directly relevant to their possible use as human food. However, for comparison of data from different sites or the same site at different times, or even between individuals at the same site, the values should be expressed on the basis of extractable lipid.

At the present time the percentage of lipid in a sample is estimated from the weight of residue remaining after evaporating the extracting solvent from an aliquot of the original extract. Hexane is not always effective in removing all lipids from the tissue sample, although organochlorine residues are efficiently extracted. It would be preferable to obtain the true lipid fraction from a separate aliquot of tissue or homogenate, using an efficient lipid solvent such as chloroform-methanol. The method of Bligh and Dyer (1959) is suggested.

## Dry-Weight Deterioration

Biological material should be dried for 24 hours at 105 to 110°C to obtain the wet-weight/dry-weight ratio. The results of the analysis can then be expressed on either a dry- or wet-weight basis.

## Limits of Detection

Current procedures permit the estimation of organochlorine concentrations in biological tissue, using an aliquot of material, down to 0.001 mg/kg wet weight for many residues, $2 \times 10^{-9}$ g/g for $p,p'$DDT and $1 \times 10^{-8}$ g/g (10 ppb) for PCBs. The latter are usually quantified by comparison with commercial mixtures such as Aroclor 1254, Phenochlor DP-5, or Clophen A.50, which give chromatograms roughly similar to those in the samples. The sums of the heights of three or more peaks are compared with the sum of the heights of the same peaks in the standard mixture. The peak interfering with that of $p,p'$DDE is not used.

At the levels of detection referred to above, a very high proportion of the values from any mussel watch program may be below the minimum detection limit. Thus, no indication can normally be obtained of trends in contamination from successive samples if the concentrations remain below this limit. The levels are considered to be well below those likely to have a biological effect on bivalves, and would be adequate for identifying sites of local or extensive pollution.

If trends in the contamination of ostensibly clean areas of coastal waters are to be studied, it will be necessary to extend the limit of detection downwards by one or perhaps two orders of magnitude. This in turn will require the concentration of extracts to very small volumes, or the use of larger tissue aliquots for extraction, or both. In any case, all contaminants will be concentrated and the chromatographic "noise" (the background response of the instrument) will be increased in proportion. The procedure will be more time-consuming (and more expensive) and should not be requested without serious consideration.

142

REFERENCES

Association of Official Analytical Chemists (1975) Official
    Methods of Analysis of the Association of Official
    Analytical Chemists, 12 ed. Washington, D.C.:
    Association of Official Analytical Chemists.
Bligh, E.G. and W.J. Dyer (1959) A rapid method of total
    lipid extraction and purification. Can. J. Biochem.
    Physiol. 37:911-917.
Dawson, R. and J.P. Riley (1977) Chlorine-containing
    pesticides and polychlorinated biphenyls in British
    coastal waters. Estuar. Coast. Mar. Sci. 4:55-69.
Food and Agricultural Organization/United Nations
    Environment Program (1976) Manual of Methods in Aquatic
    Environmental Research: Part 3, Sampling and Analysis of
    Biological Materials, edited by M. Bernhard. FAO Fish.
    Tech. Paper 158. Rome: Food and Agricultural
    Organization.
Goldberg, E.D. (1976) Strategies for Marine Pollution
    Monitoring. New York: Wiley Interscience.
Harvey, G.R. and W.G. Steinhauer (1976) Transport pathways
    of polychlorinated biphenyls in Atlantic water. J. Mar.
    Res. 34(4):561-575.
Holden, A.V. (1973) Mercury and organochlorine residue
    analysis of fish and aquatic mammals. Pestic. Sci.
    4:399-408.
Musty, P.R. and G. Nickless (1974) Use of Amerlite XAD-4 for
    extraction and recovery of chlorinated insecticides and
    polychlorinated biphenyls from water. J. Chromatogr.
    89:185-190.
Uthe, J.F., J. Reinke, and H. O'Brodovich (1974) Field
    studies on the use of coated porous polyurethane plugs
    as indwelling monitors of organochlorine pesticide and
    polychlorinated biphenyl contents of streams. Environ.
    Lett. 6(2):103-115.
Wells, D.E. and S.J. Johnstone (1977) Method for the
    separation of organochlorine residues before gas-liquid
    chromatographic analysis. J. Chromatogr. 140:17-28.
Wells, D.E. and S.J. Johnstone (1978) The occurrence of
    organochlorine residues in rainwater. Water Air Soil
    Pollut. 9:271-280.

5

RADIONUCLIDES


EDWARD D. GOLDBERG (Chairman), Scripps Institution of
    Oceanography, La Jolla, California
SCOTT FOWLER, International Atomic Energy Agency, Monaco
R. FUKAI, International Laboratory of Marine Radioactivity,
    Monaco


INTRODUCTION

The radionuclide data from the first year of the U.S.
Mussel Watch Program present some intriguing problems that
should be addressed in order to further evaluate the
usefulness of bivalves as sentinels for measuring
environmental levels of transuranics and other important
artificially produced radionuclides in future monitoring
programs.  One of the principal aims of using bivalve
indicators is to gain information on relative levels of
pollutants in the surrounding environment, including the
location of pollutant hot spots.  For example, the finding
of two- to three-fold higher levels of $^{239,240}$Pu in mussels
from an area near a nuclear reactor might signal
contamination originating from the reactor effluent.
However, the lack of additional samples, the presence of
normal levels of $^{238}$Pu and $^{137}$Cs at the same site, and some
evidence that mussels of the same species may accumulate a
given transuranic in differing degrees limit drawing any
definite conclusion about the origin of the plutonium.
Likewise, the increased ratios of $^{241}$Am to $^{239,240}$Pu in
samples from the West Coast compared to those from the East
Coast are difficult to explain because of the lack of
ancillary data needed to test any hypothesis proposed to
account for the set of observations.  Additional sampling
coupled with a modest research effort on the behavior of
transuranics in bivalves would enhance our interpretation of
the results from present programs and aid in assessing

143

strategies for future radionuclide surveys using these organisms.

It has been suggested that the presence of $^{244}$Cm in two mussel samples might indicate a local special source of transuranic contamination. Because of the lack of intense sampling at these sites, it is difficult to verify this hypothesis. Sediment samples should be taken in the vicinity of the mussel beds and the transuranic profile should be examined. In addition, analyzing invertebrates such as starfish, which are known to accumulate certain transuranics to higher levels than mussels, might also be of use in confirming the presence of this radionuclide.

Much lower ratios of $^{137}$Cs to $^{239,240}$Pu in mussel tissues than are normally found in the water or sediment imply enhanced bioavailability of $^{239,240}$Pu compared with $^{237}$Cs. The same result might also be a function of the retention time of cesium and plutonium if, in fact, cesium is lost from mussels much more rapidly than plutonium. Simultaneous uptake and loss experiments using two gamma emitters, $^{237}$Pu and $^{137}$Cs, would shed some light on this question.

Certainly the most intriguing aspect of the U.S. Mussel Watch survey was the observation that West Coast samples generally exhibited higher concentrations of plutonium and much higher $^{241}$Am to $^{239,240}$Pu ratios than were found along the Gulf and East Coasts. From the current data, it is difficult to conclude if these observations represent different transuranic concentrations at the source, different biological availabilities of transuranics at the source, or different specific bioconcentrating abilities of the mussels examined.

The first step in resolving this question is to sample transuranics in water (soluble and particulate phases) and sediments at the same sites as the mussels. Particulates and sediments are of particular interest, since there is some evidence from the U.S. Mussel Watch data that these components may be the most important source of transuranics for mussels.

If it can be established that the concentrations of transuranics at the source are similar on both the East and West Coasts, then mussel transplantation experiments should be initiated to test the hypothesis of enhanced bioavailability of West Coast transuranics. Extreme care should be taken in these studies not to unduly stress the organisms. For example, spawning can be induced in mussels that undergo rapid temperature changes.

The third possibility, that the two species of mussels, Mytilus californianus and M. edulis, might display different abilities to bioaccumulate transuranics, should also be examined in detail. Simultaneous radiotracer experiments using $^{237}Pu$ and $^{241}Am$ would be particularly useful for this purpose. Both M. californianus and M. edulis from the same site could be exposed to various ratios of two radionuclides and their uptake followed by means of gamma counting.

In addition to radiotracer studies, it would be instructive to examine the tissue distribution of $^{241}Am$ and $^{239,240}Pu$ in both species of Mytilus from the West Coast as well as in the same species (M. edulis) from both coasts. The fact that the ratios of $^{241}Am$ to $^{239,240}Pu$ in byssus differ markedly from those in whole soft parts from the same individuals suggests that certain tissues in mussels display different affinities for transuranics. Furthermore, unpublished data from panel member Fowler and co-workers show that the gonads of Mediterranean starfish display enhanced ratios of $^{241}Am$ to $^{239,240}Pu$ (an increase of from 2 to 30 during the period when the animals are preparing to spawn). If similar biological fractionation processes occur in mussels, it is possible that mature gonadal tissue may contribute significantly to the very high ratios of $^{241}Am$ to $^{239,240}Pu$ observed in the total soft parts of certain mussels.

Seasonal trends in transuranic concentration at a given site have suggested that the biological half-times for these radionuclides may be relatively short, on the order of weeks to months. It also appeared that plutonium isotopes were lost at a more rapid rate than $^{241}Am$.

Growth measurements were not made over the duration of the study, so it is difficult to determine if the observed decreases in transuranic concentration were due to the real elimination or were merely a result of new tissue added during growth which contained low transuranic concentration. It is known that rapid growth can strongly affect the turnover time of transuranics in mussels. Fowler and others (1975) and Guary and Fowler (in press) have shown in laboratory and field studies that biological half-times for plutonium in the slowly exchanging compartment of nongrowing mussels are on the order of 2 years. However, in actively growing individuals, the half-time decreases to about 7 months. These studies have also revealed that in actively growing mussels plutonium is turned over roughly 2.5 times more rapidly than americium. This may partially explain why concentrations of $^{239,240}Pu$ dropped more rapidly than those of $^{241}Am$ in mussel samples from Bodega Head.

The results of Guary and Fowler (in press) indicate that
loss of $^{241}$Am and plutonium from mussels can best be
represented by a two-compartment model. For both
radionuclides, it appears that a substantial fraction of the
accumulated contaminants (70 percent) is eliminated from the
two fast compartments of mussels with half-times on the
order of 1 to 3 weeks. If the decrease in transuranics
noted in the Bodega Head samples represents a real excretion
process, it is possible that the decline is taking place
from the rapidly exchanging compartment only, and that these
rates may not be expected to continue over longer time
intervals. In any event, because of the relatively rapid
transuranic turnover associated with a substantial fraction
of the animal's body burden, sampling in the areas where
episodic inputs are expected would have to be carried out at
intervals commensurate with the biological half-lives in the
rapidly exchanging compartment.

The biological half-life of a given transuranic will be
a function of the mussel's metabolic state. This, in turn,
suggests that the biological half-life will be related to
mussel age. More data are needed on the degree of
transuranic retention in individuals of different ages.
This type of information would be particularly useful in any
attempt to determine the age of mussels by means of their
radionuclide content (Griffin and others 1980).

Considering that most transuranic monitoring programs
will center around sites where episodic events take place,
some effort should be spent in examining the radionuclide
turnover times in field populations of mussels that have
been contaminated in situ.

## ANALYTICAL QUALITY CONTROL

As low-level measurements of the radionuclides in
question--especially of transuranics $^{238}$Pu, $^{239,240}$Pu,
$^{241}$Am, etc.--require rather sophisticated separation and
clean-up techniques, it is considered essential that
intralaboratory as well as interlaboratory comparability of
the data of these measurements is ensured by appropriate
analytical quality control. Intralaboratory data quality
control can be achieved by repeated analyses of standard
reference materials or reference samples of known
radionuclide concentrations. Interlaboratory comparability
of data should be ensured by organizing intercalibration
exercises on homogeneous samples of unknown radionuclide
concentrations. After completion of the intercalibration

exercises, the samples employed for these exercises can be designated as "reference samples," in which concentrations of the radionuclides concerned are already known with sufficient accuracy, and can be used for the purpose of intralaboratory analytical quality control. It is preferable that these samples be naturally contaminated rather than artificially spiked, in order to avoid any difference of chemical forms of the radionuclides.

Several international intercalibration exercises for radionuclide measurements, including those of transuranics, have been organized by the IAEA during past years (Table 5.1). All samples distributed to date in these intercalibration exercises have been those contaminated in situ with radionuclides at relatively high levels. The table includes the results of the determinations of transuranic elements reported by individual laboratories on the Monaco Laboratory clam sample (code no. MA-B-1), the Aplysia sample (code no. MA-B-2), and the marine sediment sample (code no. SD-B-3). The results of similar attempts made among the institutions in the United States by the Environmental Measurement Laboratory of the U.S. Department of Energy are given in Tables 5.2, 5.3, and 5.4.

The comparability of $^{238}$Pu and $^{239,240}$Pu measurements on these environmental samples among experienced laboratories is generally satisfactory, the agreement of results being within a range of approximately 10 to 20 percent. However, only a limited number of laboratories have reported the results of $^{241}$Am measurements. This demonstrates that, while the analytical methods for the measurements of plutonium isotopes have been established, methods for americium and cesium determinations are still in a developing stage and only a small number of laboratories are capable of producing reliable data. It should be noted, however, that the data of this small number of laboratories are in fairly good agreement with each other, despite the different separation procedures employed by individual laboratories.

As the preparation of standard reference materials is an extremely expensive endeavor and the availability of reference samples is limited, it is recommended that materials already available should be used as effectively as possible. Reference samples are available from the IAEA's Monaco Laboratory (Clam sample [MA-B-1], Aplysia sample [MA-B-2], and marine sediment sample [SD-B-3]). These samples contain relatively high levels of transuranic nuclides and determinations of transuranics can be carried out by using relatively small samples (a few grams). This means that the

TABLE 5.1  Results of Intercalibration Exercises on Transuranic Measurements in Marine Environmental Samples

| Lab. No. | MAB-1 (1976) | | | MA-B-2 (1976) | | |
|---|---|---|---|---|---|---|
| | $238$Pu[a] | $239+240$Pu[a] | $241$Am[a] | $238$Pu[a] | $239+240$Pu[a] | $241$Am[a] |
| 8 | < 4 | 50 ±20 | — | <18 | <12 | — |
| 10 | <44 | <25 | — | <26 | <11 | — |
| 11 | — | 60 ±10 | — | — | 20 ±5 | — |
| 13 | — | 34.3±15 | — | — | 2.45±0.98 | — |
| 14 | — | <60 | — | — | <70 | — |
| 17 | 1.41±0.25 | 38.3± 1.4 | 18.6±2.5 | 0.76±0.12 | 1.05±0.17 | 0.88±0.35 |
| 20 | 2.6 ±0.8 | 43 ± 3 | — | — | — | — |
| 21 | — | — | 35 ±2 | — | — | — |
| 31 | 1.5 ±0.2 | 36 ± 4 | 15 ±2 | 0.95±0.06 | 1.0 ±0.1 | 0.8 ±0.2 |

| Lab. No. | SD-B-3 (1978) | | |
|---|---|---|---|
| | $238$Pu[a] | $239+240$Pu[a] | $241$Am[a] |
| 7' | 25 ± 3 | 479± 7 | 163± 7 |
| 8' | 22.3±1.4 | 659± 14 | 210± 6.3 |
| 15' | 60 ±20 | 1380± 90 | — |
| 22' | 20 ± 1 | 600± 40 | 100± 6 |
| 23' | — | — | 120± 30[b] |
| | — | — | 700±500[b] |
| 25' | 23 ± 3 | 666± 64 | 167± 15 |
| 30' | 40 ±10 | 610± 30 | — |
| 33' | 22 ± 2 | 670± 40 | — |
| 35' | 23 ± 8 | 730± 51 | — |
| 40' | — | 60± 4 | — |
| 42' | <230 | 680±140 | — |
| 48' | 25 ± 7 | 660± 30 | 160± 20 |

[a] Expressed in fCi/g-dried matter
[b] Measured by $\gamma$-spectrometry

SOURCE: S. Fowler and R. Fukai, panel members, unpublished data.

TABLE 5.2   Results of the $^{238}$Pu Intercomparison

| Laboratory | FW | SW | RS | MS | SF | VM | SO |
|---|---|---|---|---|---|---|---|
| A. *Data in pCi/kg* | | | | | | | |
| ANL | .026 | .029 | 1.4 | 22. | 1.5 | 4.5 | 58. |
| LLL-N | – | .03 | – | 20. | 1.7 | 4.1 | – |
| SIO | .024 | .027 | 1.1 | 17. | 1.9 | – | – |
| ORNL-B | – | – | – | – | – | 4.3 | 58. |
| BARC | – | .02 | .58 | 9.6 | 3.3 | 2.1 | 28. |
| WHOI | .021 | .026 | 1.3 | 19. | 1.4 | 2.9 | 51. |
| OSU | .030 | .023 | 1.5 | 23. | 1.4 | – | – |
| UW | .053 | .101 | 7. | 14. | 6. | 18. | 32. |
| LDGO | – | – | 1.4 | 20. | – | – | – |
| TAM | .029 | .027 | 1.4 | 21. | – | – | 53. |
| YU | – | – | – | 18. | – | – | – |
| SRL | .013 | .027 | 1.4 | 15. | 1.5 | 3.5 | 44. |
| PNL | – | – | – | – | – | 4.5 | 63. |
| MOAL[a] | .026 | .027 | 1.4 | 19. | 1.6 | 4.2 | 52. |
| B. *Ratio of Each Laboratory to the MOAL* | | | | | | | |
| ANL | 1.00 | 1.07 | 1.00 | 1.16 | .94 | 1.07 | 1.12 |
| LLL-N | – | 1.11 | – | 1.05 | 1.06 | .98 | – |
| SIO | .93 | 1.00 | .78 | .90 | 1.19 | – | – |
| ORNL-B | – | – | – | – | – | 1.02 | 1.12 |
| BARC | – | .74 | .41 | .51 | 2.02 | .51 | .54 |
| WHOI | .81 | .85 | .96 | 1.00 | .87 | .69 | .98 |
| OSU | 1.16 | .85 | 1.03 | 1.21 | .95 | – | – |
| UW | 2.04 | 3.75 | 5.0 | .74 | 3.8 | 4.3 | .62 |
| LDGO | – | – | 1.00 | 1.05 | – | – | – |
| TAM | 1.11 | 1.00 | 1.00 | 1.11 | – | – | 1.02 |
| YU | – | – | – | .95 | – | – | – |
| SRL | .50 | 1.00 | .99 | .80 | .95 | .83 | .85 |
| PNL | – | – | – | – | – | 1.07 | 1.21 |

[a]MOAL = Median of all laboratories.

SOURCE: Volchok and Feiner (1979).

total amounts of stable lanthanides introduced into the
analytical system are limited, so that special procedures
for separating lanthanides completely from americium are not
necessarily required.  However, the ability to perform
americium measurements on these relatively high-level
samples with a given procedure does not necessarily
guarantee the applicability of the procedure to lower-level
samples.  Reference samples with lower levels of

150

TABLE 5.3    Results of the $^{239}$Pu Intercomparison

| Laboratory | FW | SW | RS | MS | SF | VM | SO |
|---|---|---|---|---|---|---|---|
| A. *Data in pCi/kg* | | | | | | | |
| ANL | .097 | .101 | 44 | 600 | 20 | 48 | 2680 |
| LLL-N | – | .107 | – | 570 | 20 | 43 | – |
| SIO | .10 | .10 | 40 | 540 | 17 | – | – |
| ORNL-B | – | – | – | – | – | 43 | 3170 |
| BARC | .80 | .081 | 94 | 210 | 15 | 13 | 1600 |
| WHOI | .094 | .094 | 36 | 560 | 18 | 36 | 2750 |
| OSU | .131 | .113 | 38 | 630 | 19 | – | – |
| UW | .035 | .114 | 21 | 370 | 13 | 26 | 1700 |
| LDGO | – | – | 35 | 530 | – | – | – |
| TAM | .063 | .090 | 35 | 586 | – | – | 2940 |
| YU | – | – | 32 | 508 | – | – | 2100 |
| SRL | .060 | .066 | 36 | 430 | 17 | 38 | 2350 |
| PNL | – | – | – | – | – | 38 | 2530 |
| MOAL[a] | .096 | .100 | 36 | 540 | 18 | 38 | 2530 |
| B. *Ratio of Each Laboratory to the MOAL* | | | | | | | |
| ANL | 1.01 | 1.01 | 1.24 | 1.12 | 1.11 | 1.26 | 1.06 |
| LLL-N | – | 1.07 | – | 1.06 | 1.11 | 1.13 | – |
| SIO | 1.04 | 1.00 | 1.11 | 1.00 | .94 | – | – |
| ORNL-B | – | – | – | – | – | 1.13 | 1.25 |
| BARC | 8.33 | .81 | 2.61 | .39 | .83 | .34 | .63 |
| WHOI | .98 | .94 | .99 | 1.04 | 1.00 | .95 | 1.09 |
| OSU | 1.37 | 1.13 | 1.05 | 1.16 | 1.06 | – | – |
| UW | .37 | 1.14 | .59 | .68 | .72 | .68 | .67 |
| LDGO | – | – | .96 | .98 | – | – | – |
| TAM | .66 | .90 | .97 | 1.09 | – | – | 1.16 |
| YU | – | – | .90 | .60 | – | – | .83 |
| SRL | .63 | .66 | 1.00 | .67 | .94 | 1.00 | .93 |
| PNL | – | – | – | – | – | 1.00 | 1.00 |

[a]MOAL = Median of all laboratories.

SOURCE: Volchok and Feiner (1979).

transuranics are required for the analytical quality control of low-level measurements. An attempt to make available such low-level reference samples is now in progress.

Analysis of mussel shells for transuranics may pose problems because of the high calcium and phosphate content of the shells. The preparation of a shell reference sample specifically for the purpose of the mussel watch is recommended.

TABLE 5.4    Results of the $^{241}$Am Intercomparison

| Laboratory | FW | SW | RS | MS | SF | VM | SO |
|---|---|---|---|---|---|---|---|
| A. *Data in pCi/kg* | | | | | | | |
| ANL | .027 | .026 | 7.0 | 140 | 4.0 | 6.2 | 250 |
| SIO | – | .025 | 10.0 | 190 | 4.7 | – | – |
| ORNL-B | – | – | – | – | – | 13. | 390 |
| BARC | .030 | .036 | – | – | 3.6 | – | 180 |
| WHOI | .037 | .037 | 8.8 | 190 | 6.1 | 13.1 | 350 |
| OSU | .040 | .039 | 9.9 | 160 | 6.0 | – | – |
| UW | 3.10 | – | – | – | – | – | 390 |
| SRL | <2[b] | <2[b] | <120[b] | <230[b] | <160[b] | <110[b] | 310 |
| PNL | – | – | – | – | – | 19. | 340 |
| MOAL[a] | .037 | .036 | 9.4 | 180 | 4.7 | 13. | 340 |
| B. *Ratio of Each Laboratory to the MOAL* | | | | | | | |
| ANL | .73 | .72 | .74 | .78 | .85 | .48 | .74 |
| SIO | – | .70 | 1.06 | 1.06 | 1.00 | – | – |
| ORNL-B | – | – | – | – | – | 1.00 | 1.15 |
| BARC | .81 | 1.00 | – | – | .77 | – | .53 |
| WHOI | 1.00 | 1.03 | .94 | 1.06 | 1.30 | 1.01 | 1.03 |
| OSU | 1.08 | 1.08 | 1.05 | .89 | 1.28 | – | – |
| UW | 8.4 | – | – | – | – | – | 1.15 |
| SRL | – | – | – | – | – | – | .91 |
| PNL | – | – | – | – | – | 1.45 | 1.00 |

[a] MOAL = Median of all laboratories.
[b] Not used in calculation of MOAL.

SOURCE: Volchok and Feiner (1979).

APPENDIX 5-A

ANALYTICAL TECHNIQUES

The analyses of the transuranics $^{238}$Pu, $^{239,240}$Pu, and $^{241}$Am include a determination step involving alpha spectrometry.  Thus, the final preparates are sought with minimum association with other materials that can absorb alpha rays or that can introduce alpha radiation.  Most methods are based upon the general schemes, which have been described by the Monaco Laboratory of the IAEA and which are depicted in Figures 5-A.1, 5-A.2, and 5-A.3 (Ballestra and others 1978).  Recipes based upon these general schemes,

used at the Woods Hole Oceanographic Institution, follow.
The analyses of $^{237}Cs$ and $^{90}Sr$ are also included in the
methodology.

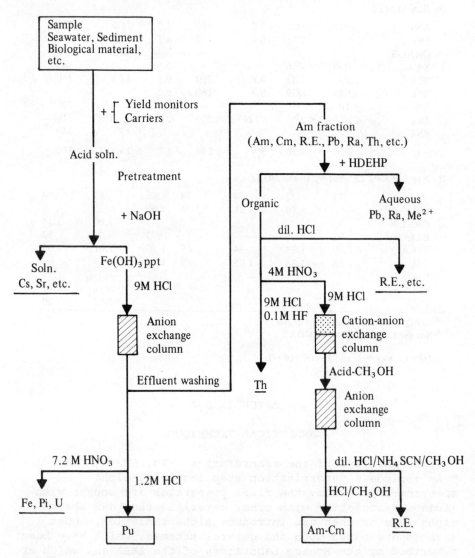

FIGURE 5-A.1  General scheme for radiochemical separation steps for the transuranic
measurements. (Reprinted with permission from Ballestra and others, Low-level determination of transuranic elements in marine environmental samples, Paper no. 15, presented at
the Symposium on the Determination of Radionuclides in Environmental and Biological
Materials, October 1978, International Atomic Energy Agency, Vienna.)

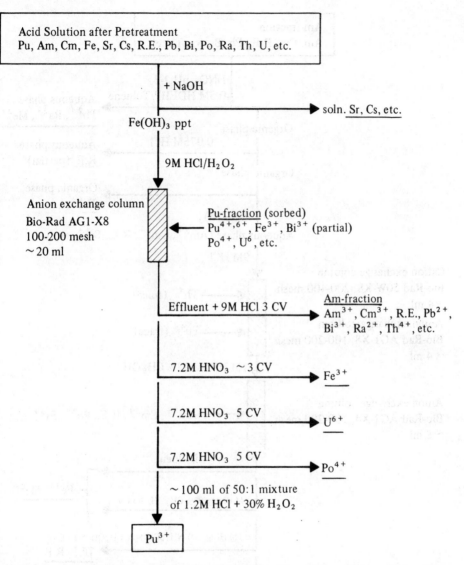

FIGURE 5-A.2 Separation scheme for plutonium. (Reprinted with permission from Ballestra and others, Low-level determination of transuranic elements in marine environmental samples, Paper no. 15, presented at the Symposium on the Determination of Radionuclides in Environmental and Biological Materials, October 1978, International Atomic Energy Agency, Vienna.)

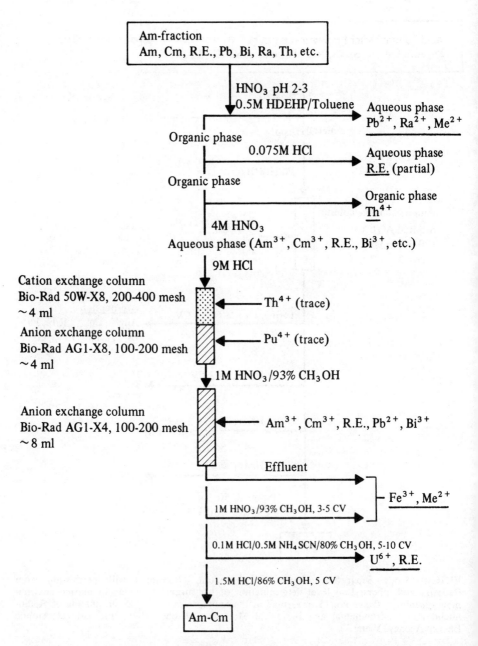

FIGURE 5-A.3 Separation scheme for americium-curium. (Reprinted with permission from Ballestra and others, Low-level determination of transuranic elements in marine environmental samples, Paper no. 15, presented at the Symposium on the Determination of Radionuclides in Environmental and Biological Materials, October 1978, International Atomic Energy Agency, Vienna.)

The Woods Hole Procedure for the Analyses
of Strontium, Cesium, Plutonium, and Americium
in Biological Samples[1]

1-1   Between 0.5 and 1 kg, wet weight, of a sample is
      dried to constant weight.
1-2   Macerate the sample well with a stirring rod.  Wet
      sample with water.  Wet sample with 8 $\underline{N}$ $HNO_3$ and note
      if $CO_2$ is given off.  Be prepared to add octyl
      alcohol dropwise to reduce bubbling if necessary.
1-3   Add 400 ml of 8 $\underline{N}$ $HNO_3$ and let stand at room
      temperature for one hour.
1-4   Add the following carriers and yield monitors:

          10 ml Sr   (100 mg Sr/ml)
           2 ml Nd   ( 25 mg Nd/ml)
           2 ml Cs   (  8 mg Cs/ml)
           1 ml Pu   (  2 dpm Pu/ml)
           1 ml Am   (  1 dpm Am/ml)

1-5   Record the carrier and yield monitor standard
      identification in notebook.
1-6   Digest sample on moderate heat until liquid is clear.
1-7   After sufficient digestion, if a precipitate remains,
      filter sample through an acid-wetted No. 42 Whatman
      filter paper and retain filter until sample analysis
      data are obtained and it is ascertained that sample
      activity has not been lost.  If the sample is to be
      analyzed for strontium, follow the strontium
      separation procedure.  Otherwise, the analysis may
      proceed to the plutonium separation procedure.


Sr-Separation

Sr-1  Concentrate the solution from Step 1-7 to 100 ml.
      Add 150 ml concentrated $HNO_3$ and reevaporate to 100
      ml or until all $Sr(NO_3)_2$ salts have formed.  Allow to
      cool.
Sr-2  Place the sample and a wash bottle of concentrated
      $HNO_3$ on ice for chilling.  Add 50 ml of fuming (90
      percent) $HNO_3$ using a glass measuring device.
Sr-3  Continue chilling for half an hour.  Label 2 flasks
      (1 l); one for Sr, the other for Pu.  Filter the
      sample with a Buchner funnel through a 934 AH glass
      fiber filter.  Wash the $Sr(NO_3)_2$ precipitate from
      beaker and wash the filter paper with cold

concentrated $HNO_3$. Save the filtrate for Pu analysis. Change filter flasks and dissolve $Sr(NO_3)_2$ with cold water, using as little as possible. Evaporate down filtrate in clean 400 ml beaker to 50 ml, and cool.

Sr-4   Add 1 ml of Fe carrier (2 mg Fe/ml), stir with glass rod, and rinse down the beaker walls with water.

Sr-5   Add concentrated $NH_4OH$ to the solution until pH is 9. An iron hydroxide precipitate will form.

Sr-6   Warm the sample 10 minutes and cool to room temperature.

Sr-7   Filter through a No. 541 Whatman filter in a glass funnel into a clean, labelled beaker. Wash beaker and filter with 10 ml of pH-8 $H_2O$. Discard any precipitate.

Sr-8   Adjust the pH of the solution to between 4 and 6 with 3 $\underline{N}$ HCl. Repeat Steps Sr-4 through Sr-7. Record date and time of the last iron scavenge.

Sr-9   Readjust the pH to 4 to 6 with 3 $\underline{N}$ HCl.

Sr-10   Add 4 ml barium acetate buffer and heat to near boiling (see note Sr-A, appended to these techniques).

Sr-11   Add 1 ml 1 M $Na_2CrO_4$ and continue heating with occasional stirring for 30 minutes.

Sr-12   Cool solution and filter precipitate on No. 42 Whatman filter paper. Wash with water and discard precipitate.

Sr-13   Repeat Steps Sr-10 to Sr-12.

Sr-14   Add 50 ml saturated $(NH_4)_2CO_3$ solution and stir 2 minutes. Remove stirrer and allow precipitate to settle.

Sr-15   Aspirate off supernate and discard.

Sr-16   Wash precipitate with 100 ml 5 percent $(NH_4)_2CO_3$; allow to settle, and aspirate off the supernate.

Sr-17   Repeat the $(NH_4)_2CO_3$ wash.

Sr-18   Dissolve the precipitate in 3 $\underline{N}$ HCl and transfer to a tared 125 ml polyethylene bottle. Write tare weight on bottle as well as the iron scavenge date.

Sr-19   Add 1 ml of Fe carrier.

Sr-20   The sample must be stored for at least 2 weeks between the iron scavenge date and the Y-90 separation. Use standard techniques for the separation and assay of Y-90.

Pu Separation

Pu-1    The digested solution from Step 1-7, or the filtrate
        from Step Sr-3, is evaporated to 100 ml and cooled.
Pu-2    Add an equal volume of 0.5 $\underline{N}$ HNO$_3$. When ready to put
        through column, add 1 g NaNO$_2$ to each 100 ml of
        solution and mix until dissolved.
Pu-3    Pass the sample through an ion exchange column at a
        flow rate of 0.5 ml/min, filtering the sample through
        a 934 AH glass fiber filter into the reservoir (see
        Note Pu-A, appended to these techniques). Collect
        the eluate in a 2 l beaker for Cs, Am, and Fe
        analysis.
Pu-4    Rinse the reservoir and column 3 times with 8 $\underline{N}$ HNO$_3$,
        collecting the washes with the original eluate.
Pu-5    Add 150 ml of 8 $\underline{N}$ HNO$_3$ through the column, collecting
        with above. Evaporate this eluate to about 100 ml
        and proceed to Step Cs-1.
Pu-6    Remove reservoir, push down glass wool, and add 2 to
        4 ml of 12 $\underline{N}$ HCl to the column. Then add the
        remaining 125 ml of 12 $\underline{N}$ HCl in the reservoir. Allow
        the HCl wash to pass through the column at a fast
        flow rate. Discard this eluate. Do not allow the
        column to stand at this point, as channeling will
        occur in the resin.
Pu-7    Make a fresh solution of 100 ml concentrated HCl plus
        5 ml 1 M NH$_4$I. Add to the column and collect eluate
        in a 250 ml beaker. Rinse column and reservoir twice
        with 25 ml concentrated HCl, collecting the wash.
Pu-8    Evaporate to dryness, adding 1 ml concentrated HNO$_3$
        prior to initial heating. As samples approach
        dryness, add about 2 ml of concentrated HNO$_3$ and
        about 2 ml of concentrated HCl. Do this 5 to 6 times
        until there is very little residue. Finally, rinse
        with a few ml of concentrated HCl.
Pu-9    Before passing the sample through the second Pu
        column (see Note Pu-B, appended to these techniques),
        add enough concentrated HNO$_3$ to cover the bottom of
        the beaker from Step Pu-8. Warm slightly.
Pu-10   Add 5 ml 8 $\underline{N}$ HNO$_3$ to dissolve the residue in the
        beaker. Add 50 mg NaNO$_2$ to the warm solution.
Pu-11   Pass the solution through the column at 0.5 ml/min.
        Rinse the beaker 3 times with a few ml of 8 $\underline{N}$ HNO$_3$,
        adding each rinse to column.
Pu-12   Wash the column with 20 ml of 8 $\underline{N}$ HNO$_3$ and 20 ml of
        concentrated HCl. These washes are discarded.

Pu-13    Add 1 ml of 1 M NH₄I to 20 ml concentrated HCl.  Stir
         and add to column at a fast rate, collecting the
         eluate in 100 ml beaker.  Rinse the column with 10 ml
         concentrated HCl.

Pu-14    Add 1 ml concentrated HNO₃ to eluate and evaporate to
         dryness.  Treat the remaining salts several times
         with equal portions of concentrated HCl and
         concentrated HNO₃.  When all volatile salts are
         sublimed, rinse the inside of the beaker with a few
         ml of concentrated HCl and evaporate to dryness.

Pu-15    Before passing the sample through the third Pu column
         (see Note Pu-C, appended to these techniques), add
         enough concentrated HCl to cover the bottom of the
         beaker.  Warm slightly and add 5 ml concentrated HCl
         and 1 drop of water H₂O₂ (30 percent).

Pu-16    Pass the solution through the column at 0.5 ml/min.
         Rinse the walls of the column 3 times with a few ml
         of concentrated HCl.  Wash the column with 20 ml
         concentrated HCl and discard eluate.

Pu-17    Elute the Pu with 2 to 10 ml portions of 1 ml 1 M
         NH₄I + 20 ml concentrated HCl.  Rinse with 2 to 5 ml
         portions of concentrated HCl, collecting in a 100 ml
         beaker.

Pu-18    Evaporate to dryness.  Add 2 ml of concentrated HCl
         and concentrated HNO₃ repeatedly as the sample
         approaches dryness.  The final rinse is with 2 to 3
         ml concentrated HCl evaporating to dryness.

Pu-19    Cover with paraffin to await electroplating.


Cs Separation

Cs-1     The eluate from Step Pu-5 is evaporated to 100 ml.
         Then the volume is topped to 1 1 with water and
         stirred on a stirring plate.

Cs-2     While stirring, adjust the pH of the solution to 1.5
         with 10 M NaOH.  Use 8 N HNO₃ if excess NaOH is
         added.  Leave the stir bars in for the next step.

Cs-3     Weigh 1.0 g of AMP into a 20 ml beaker and add enough
         water to form a slurry.  Add the slurry to the
         sample, using water to rinse all the AMP into the
         sample beaker.  Stir 10 minutes on a stirrer to
         ensure complete mixing.  Remove stirring bar and
         allow the precipitate to settle overnight.

Cs-4     Decant the supernate and save for americium analysis.
         Rinse the Cs-AMP precipitate with 0.02 N HCl into a
         centrifuge tube.

Cs-5    Spin down Cs-AMP; pour supernate into Am solution.
        The Cs-AMP is resuspended in 0.02 $\underline{N}$ HCl until the
        washings are colorless.  Add the rinses to the Am
        sample.
Cs-6    Dissolve the AMP-Cs in 20 ml 0.75 M NaOH.  Mix well
        to dissolve Cs-AMP.  If yellow AMP is still present,
        add 10 M NaOH dropwise to dissolve the AMP.
        Centrifuge the dissolved AMP to remove any hydroxide
        precipitate that may be present.  Add the precipitate
        to the Am fraction.
Cs-7    Add 2 ml of 20 percent EDTA to the supernate.


Column Cesium Procedure

Cs-8    Add the sample to a previously prepared column (see
        Note Cs-A appended to these techniques), rinsing the
        tube 3 times with a small amount of water.
Cs-9    Wash the column with 375 ml of 0.3 $\underline{N}$ HCl.  This
        fraction is discarded.
Cs-10   Elute the Cs fraction with 65 ml of 3 $\underline{N}$ HCl,
        collecting it in a 150 ml beaker.
Cs-11   Evaporate solution on a hot plate to dryness.
Cs-12   Add 1 ml of 10 M NaOH to the dry salt.  Add 10 ml
        water in several aliquots to transfer the sample to a
        centrifuge tube, being careful to remove all salts
        from the bottom of the beaker.  If a precipitate is
        present, pour the sample through a Whatman filter
        into a centrifuge tube.  Always saturate the filter
        with water first.
Cs-13   Heat the sample at 80°C in a water bath for 10
        minutes.  Then allow it to cool to room temperature.
Cs-14   Centrifuge to remove any hydroxide precipitate that
        may have formed.  Transfer the supernate to a clean
        centrifuge tube and discard precipitate.
Cs-15   Heat at 80°C for 10 minutes.
Cs-16   Add 2 ml 10 percent $H_2PtCl_6$ (chloroplatinic acid)
        while hot.
Cs-17   Continue heating for 30 minutes.
Cs-18   Cool in refrigerator for 30 minutes.
Cs-19   Heat at 80°C for 10 minutes.
Cs-20   Refrigerate overnight.  The Cs precipitate is ready
        for mounting.

## Cs Mounting Procedure for Beta Counting

Cs-21 Filter preparations:
Rinse a Millipore filter HA (25 mm diameter) in 3.0 $\underline{N}$
HCl. Rinse in water. Dry at 60°C for at least half
an hour and weigh when cool.

Cs-22 Place filter on center of a Schleicher & Schuell
filter apparatus. Wet with water and attach chimney.

Cs-23 Turn on vacuum and overlay the chilled sample at a
depth of 1-2 mm onto the filter with Pasteur pipet.
Rinse the tube with cold water and add to filter in
the same manner. Avoid adding too much at once, as
the precipitate will cling to the chimney wall.
Remove chimney with vacuum still on.

Cs-24 Dry the filtered sample in oven at 60°C for about
half an hour, cool and weigh.

Cs-25 Mount filter on a Plexiglas sheet 5.6 cm × 10.2 cm ×
0.16 cm thick and place Mylar film over the filter
without disturbing the filter surface.


## Am Separation

Am-1 Place Cs-AMP supernate from Step Cs-5 on a stirring
hot plate and warm.

Am-2 Add 50 ml concentrated $NH_4OH$ while stirring. Stir
for 10 minutes.

Am-3 Remove stir bar, cool and cover with Saran Wrap.
Allow the precipitate to settle overnight.

Am-4 Aspirate off supernate and discard.

Am-5 Rinse precipitate into a 250 ml centrifuge bottle
with pH-8 water (few drops concentrated $NH_4OH$/500 ml
water).

Am-6 Centrifuge for 15 minutes.

Am-7 Discard supernate.

Am-8 Dissolve precipitate in minimal 8 $\underline{N}$ $HNO_3$ and dilute
to 450 ml.

Am-9 Add 450 ml of saturated oxalic acid and adjust the pH
carefully to 1.0 with concentrated $NH_4OH$. Allow the
precipitate to settle overnight.

Am-10 The americium is treated in standard ways to produce
an americium electroplate separate.

## Notes

1. Prepared by J.M. Palmieri, H.L. Quinby, and D.R. Mann.

Sr-A: The barium acetate buffer is prepared by combining 9
   g BaCl, 40 ml concentrated acetic acid, and 230 g
   ammonium acetate and diluting to 1 l with distilled
   water.

Pu-A: The first Pu column dimension is 12 mm ID × 150 mm
   long, containing 25 ml of wet settled Bio-Rad AG 21K (50
   to 100 mesh) in chloride form. The column is
   preconditioned with 50 ml of concentrated $HNO_3$, followed
   by 100 ml of 8 N $HNO_3$ to which 1 g of $NaNO_2$ is added.

Pu-B: The second Pu column dimension is 10 mm ID × 50 mm
   long, containing 2 ml of wet settled Bio-Rad AG 21K (50
   to 100 mesh) in chloride form. The column is
   preconditioned with 4 ml of concentrated $HNO_3$, followed
   by 20 ml of 8 N $HNO_3$ followed by 20 ml of 8 N $HNO_3$ to
   which 0.2 g $NaNO_2$ has been added.

Pu-C: The third Pu column is prepared in the same manner as
   the second Pu column. The preconditioning, however, is
   with 20 ml concentrated HCl and 2 drops of 30 percent
   $H_2O_2$.

Cs-A: The column dimension is 10 mm ID × 200 mm long,
   containing 17 ml of wet, settled Bio-Rex 40 (50 to 100
   mesh) in H⁺ form. To regenerate the column, pass 10 ml
   of water through the column, followed by 200 ml of 5
   percent NaCl, then 100 ml of water. Allow time for
   column to drain between additions.

## REFERENCES

Ballestra, S., E. Holm, and R. Fukai (1978) Low-level
   determination of transuranic elements in marine
   environmental samples. Paper No. 15, presented at the
   Symposium on the Determination of Radionuclides in
   Environmental and Biological Materials, London, October
   1978, Radioactivity in the Sea Series. Vienna:
   International Atomic Energy Agency.

Fowler, S., M. Heyrand, and T.M. Beasley (1975) Experimental
   studies on plutonium kinetics in marine biota. Pages
   157-177, Impacts of Nuclear Releases into the Aquatic
   Environment. Vienna: International Atomic Energy Agency.

162

Griffin, J.J., M. Koide, V. Hodge, and E.D. Goldberg (1980)
    Determining the ages of mussels by chemical and physical
    methods. <u>In</u> Isotope Marine Chemistry. 70th Anniversary
    Volume for Y. Miyaka, Meteorological Institute. Tokyo
    (in press).

Guary, J.C. and S.W. Fowler (1979) Elimination and
    repartition du $^{241}$Am et du $^{237}$Pu chez du moule <u>Mytilus</u>
    <u>galloprovincialis</u> dans son environnement naturel. Rapp.
    Comm. Int. Mer Medit. 25 (in press).

Volchok, H.L. and M. Feiner (1979) A Radioanalytical
    Laboratory Intercomparison Exercise, October 2, 1979.
    Environmental Measurements Laboratory. New York: U.S.
    Department of Energy.

6

## MUSSEL HEALTH

BRIAN L. BAYNE (Chairman), National Environmental Research
   Council, Plymouth, Great Britain
DAVID A. BROWN, University of British Columbia, Vancouver,
   British Columbia, Canada
FLORENCE HARRISON, Lawrence Livermore Laboratories,
   Livermore, California
PAUL D. YEVICH, Environmental Protection Agency,
   Narragansett, Rhode Island

With contributions from:

GREGORIO VARELA, Universidad Central Ciudad, Madrid, Spain
DAVID L. WEBBER, University College of Swansea, Wales, Great
   Britain

## INTRODUCTION

The panel on mussel health took as its main objective
the review of procedures that are currently available for
measuring biological effects of pollutants on mussels and
other bivalves. Consequently, the primary emphasis of this
chapter is the description and assessment of such
procedures. However, in drawing attention to the monitoring
of animal health, it is necessary to establish criteria for
identifying effects of potential pollutants; accordingly
some fundamental considerations that are important in
assessing the biological impact of a pollutant are also
discussed.

One particular aspect of the fate of contaminants within
the body, and one that we believe is crucial to meaningful
study of biological effects, is the fact that many
mechanisms exist for detoxifying chemical insults. The
study of detoxification systems--and particularly of the
maximum capacity of such systems to cope with specific

163

contaminants--is at an early stage, and we are not yet in a position to generalize or to suggest particular techniques that could be applied in monitoring. Nevertheless, problems of detoxification are so central to our main theme of mussel health that we have devoted one section of this chapter to that topic, with recommendations for further research.

Measurement of biological effects imposes certain constraints on sampling procedure, and some of these are identified and discussed.

In the main body of this chapter we discuss various indices of biological effects and procedures for measurement that may be useful in mussel watch. Some of these are now reasonably well established and proven and may therefore be incorporated directly into surveillance programs. Others require further development and assessment in the field (as opposed to their use in laboratory experiments); we have tried to discriminate between these in the text and in our recommendations.

Why should we wish to measure the biological effects of marine pollutants? Assuming that the rationale is convincing, why should we use mussels as indicator organisms?

In answer to the first question, it is important to learn of effects in order to interpret the true significance of the chemical findings. Determining the concentration or body burden of contaminants is a chemical problem, but the significance in terms of environmental impact is a biological one. The health of mussels is important because of the presumption that if environmental conditions are severe enough to cause biological changes in mussels, they are likely also to have an effect on other organisms more sensitive to environmental stresses. As for the second question, the use of mussels as biological indicators can be justified as follows. Although mussels, and other estuarine bivalves, may be considered "tolerant" of a wide range of environmental stressors, the evidence suggests that this tolerance is due to an increased capacity to endure sublethal stress, during which there may be considerable disturbance to physiological and biochemical processes without inducing a lethal condition. When physiological mechanisms are stretched beyond their normal adaptive range, sublethal conditions of stress become measurable. Mussels have a wide tolerance of disturbed physiological (and biochemical) states and therefore qualify as better sentinel organisms than species with narrower (and therefore less easily measured) sublethal ranges of tolerance.

## CRITERIA FOR ESTABLISHING A BIOLOGICAL EFFECT

Consideration of the health of mussels (or of any organism) requires a discussion and understanding of stress. Selye (1950) first defined stress (as applied to mammals) as "a state of nonspecific tension in living matter." Selye identified certain hormonal, biochemical, and morphological changes with which he characterized a general stress (or adaptation) syndrome. In applying some of these ideas to fish, but recognizing that poikilothermic organisms might respond differently than homiotherms, Brett (1958) emphasized that one aspect of stress was that it "extends the adaptive responses of an animal beyond the normal range." This is an important idea, implying that it is only when an animal's evolved adaptability is exceeded that the damaging effects of stress become apparent. Bayne (1975) adapted Brett's definition for studies of bivalves, and emphasized, as had Brett, the need for a quantitative statement of stress. His definition was that "stress is a measurable alteration of a physiological (or other) steady state which is induced by an environmental change and which renders the individual (or the population) more vulnerable to further environmental change."

Two aspects of these concepts of stress are of particular significance. First, in order to establish that a stress has occurred, an altered steady-state condition should be established; this is discussed in more detail below. The second aspect of importance is that damage to the individual or to the population must be shown to result from the change. Some physiological changes that might occur as the result of man's impact on the environment may be to the animal's advantage, and we would not then consider that contamination had detrimental effects. However, if the organism's chances of survival, or its reproductive capability, or its capacity to resist further environmental change were reduced, pollution, as opposed to mere contamination, could be said to have occurred.

The requirement that damage be shown as a result of the environmental insult affects the choice of biological monitors. For example, many enzymes are extremely sensitive, under in-vitro assay conditions, to increased metal concentrations, but (quite apart from questions of the in-vivo relevance of such tests) this property does not qualify these enzymes as useful monitors of biological effects unless depressed activity of the particular enzyme can be shown (or convincingly argued) to be damaging to the growth or survival of the individual. In laboratory studies with mussels, various indices have been shown to correlate with reduced fecundity or depressed growth efficiency (Bayne and others 1978), and such indices can therefore be adopted with some confidence as measures of the real and damaging consequences of environmental deterioration.

The concept of a general stress response has been conceived and developed by Selye from his work with mammals (and mainly humans). The concept postulates a general stress syndrome that is characteristic of the body's response to a variety of environmental insults. The idea of a general physiological and biochemical response to environmental deterioration is equally applicable to studies on invertebrates. The components of the stress response may vary between species, but the paradigm of a general stress response is probably universal.

We will, later in this report, describe in greater detail some components of what is considered the general stress syndrome in mussels. In response to a change in the environment, damaging or benign, physiological mechanisms show a three-stage change in rate: there is an initial response that often involves overshoot or undershoot phenomena, a period of stabilization of the response during which changes of rate occur, and finally a new steady state is established that may or may not be measurably different from the value of the steady state that existed prior to the environmental stimulus. If the new steady state is the same as the pre-existing conditions, complete physiological adaptation, or acclimation, is said to occur; if it is measurably different from the former condition, a physiological effect is registered. However, as has been made clear, only if the physiological change can be said to be damaging to the individual in terms of its chances of survival or of its fitness can we conclude that a stress condition results.

Physiological responses can best be used to assess the mussels' health if they are integrated into the calculation of such indices as the scope for growth or the oxyen/nitrogen ratio. When changes in feeding rate, absorption efficiency, and excretion and respiration rates are all considered in terms of the balanced energy equation, a measure of the growth potential is possible. When growth potential, or the scope for growth, is reduced, we can identify a stress effect; indeed a decline in the scope for growth is the fundamental characteristic of the stress syndrome.

Other elements of the stress response have been identified, and some of these provide the most immediately useful means of measuring the effects of environmental impact. For example, changes in the stability of the lysosomal membrane constitute one element of a general cytotoxic response. As discussed in more detail later, lysosomes accumulate many of the contaminants that are taken into the body, and under certain conditions this uptake may result in weakening of the membrane binding of the lysosomal enzymes and in increased permeability of the membrane to substrate molecules. The result may be accelerated autolysis of cellular material, with consequent deterioration in the health of the individual.

The concept of the general stress syndrome implies causative links between some of the characteristic responses that constitute the syndrome; this is true in mussels. For example, it has been shown that latency of lysosomal hydrolases correlates well with the scope for growth in mussels subjected to different levels of environmental pollutants (Bayne and others 1976). These characteristic responses provide a powerful tool for assessing the biological effects of pollution. Because they constitute a general response to a variety of different stressors, they are useful in measuring the health of the individual and provide a scale of stress effects for the assessment of condition and should be used in mussel watch monitoring programs.

However, having deduced that a particular population of mussels is in poor (or very poor, or indifferent) health, we need to establish likely causative agents; is the decline in condition due to pollution, and if so, to what class of pollutant? In other words, we need specific indices that are responsive to particular classes of contaminants in addition to indices of the general stress response. There are certain specific cytological responses to certain pollutants. For example, (a) the concentration of metallothioneins in the cell may provide an index of response to particular toxic metals (Brown and others 1977a), and (b) the level of metabolic activity in the cytochrome P-450-linked, mixed-function oxidase system may provide an indication of pollution by xenobiotic organic compounds (Bend and others 1977). We recommend that some of these indices be included in mussel watch and we urge that research directed toward discovering other specific response indices be encouraged.

In summary, we can identify certain biological effects of pollution that together constitute a general stress syndrome. To a large extent these effects can be monitored as "general stress indices" and we recommend that these be incorporated in mussel watch programs as soon as possible. We also identify the need for "specific stress indices" that will help to identify causative agents of decline in mussel health.

## SAMPLING CRITERIA FOR BIOLOGICAL EFFECTS

### Seasonality

The occurrence of seasonal cycles is an important factor in the physiological, cytological, or biochemical condition of bivalve molluscs. The fundamental cycle involves reproductive condition; most bivalves breed only at particular times of the year, and in response to the considerable energy demands of gametogenesis, stores of nutrient reserves are often built up in the tissues (notably the digestive gland and the connective tissue of the mantle) during periods of nutrient availability. The nutrient reserves are then depleted as the material is used for the synthesis of lipids and proteins in the gametes. The seasonality in nutrient storage and reproductive condition takes different forms in different species, but in all species it results in seasonal changes in body weight, biochemical composition, and physiological performance. Considerable morphological changes accompany the seasonal cycles and, of particular concern in the present context, after the discharge of gametes (spawning) the individuals are often in very poor physiological and cytological condition.

The timing of sampling is therefore important. We recommend that, for the assessment of mussel health, samples be taken prior to spawning. At that time the animal's physiology is at its most stable and the results of physiological measurements are most easily interpreted. In addition, animals with ripe gametes in the reproductive follicles afford certain means for pathological assessment [such as appearance of the reproductive follicles (see Bayne and others 1978) and parasitic infestation of gametes]. Ripe individuals may also be used in the laboratory to measure fecundity and egg quality, which are important indices of physiological fitness.

A related aspect of the problem of seasonality is the relationship between nutritional status and the concentration of contaminants in the body, since bivalves in many populations experience periods of relative starvation during certain times of the year. The nutritive condition

of the animal will have some influence on toxicity of
certain pollutants. Lipid-soluble contaminants are
concentrated in fat stores; normally, turnover of these
stores is slow. During starvation, however, lipids may be
mobilized and lipid-soluble contaminants may be released in
the body.

Even under normal conditions, however, there is a
glycogen storage cycle in some bivalves (Gabbott 1976) that
alternates between high glycogen levels prior to
gametogenesis and high lipid levels when gametogenesis is
active. The situation must influence the dynamics of the
lipid-soluble contaminants, but little is known of the
effect. The considerable biochemical changes that accompany
the storge cycle should be borne in mind when sampling is
planned.

We recommend, therefore, that certain simple procedures
be employed routinely in mussel watch to record the seasonal
cycle of particular populations. In a subsequent section we
describe procedures for assessing the "condition index" of
individuals, and we demonstrate some of the seasonal changes
encountered in this simple index of body size. In some
populations a sharp decline in condition index can indicate
a spawning period; prespawned animals have a high condition
index. Quantitative cytological techniques are available,
using the procedures of stereology to assess reproductive
condition (Elias and others 1971). When such data are
available, interpretation both of the chemical
concentrations and of the biological effects is made much
easier.


## Size

Problems associated with animal size are characteristic
of sampling programs designed to determine the body burden
of pollutants. Some of these problems apply also to
biological effects measurements, for example, the possible
relationship beween size and concentrations of free amino
acids. In general, however, animal size is not a variable
of major significance in biological monitoring (e.g.,
histopathological studies, genetic indices, energy charge),
with the major exception that the rates of most
physiological processes are very dependent indeed on
individual size (Bayne 1976). Rates of oxygen consumption
and rates of filtration vary as power functions of body size
with exponents according to location, species, time of year,
salinity, temperature, suspended particulate load, and so

on.  When physiological assessments are made, animals of
different sizes must be examined and the relationship of the
physiological response to size established.  An example of
one procedure is discussed in Appendix 6-A.


## Sample Size

The sample-size requirements for monitoring biological
effects coincide to a large extent with those for other
monitoring activities.  For most aspects of
histopathological assessment and for biochemical
measurement, 25 can be considered an optimal sample size.
Physiological measurements should ideally include at least
15 individuals of different sizes.

An important aspect of the question of sample size that
is not discussed here relates to the optimal sampling
strategy in areas with patchy distributions of mussels, or
in areas subjected to complex changes in environmental
conditions over time.  We recommend only that due
consideration be given to this problem in the planning
stages of any monitoring program.


## Species Differences

There are, of course, considerable differences between
species used for most of the biological measurements
considered in this report.  Some of the differences are
indicated and discussed in the relevant sections and we
point here only to some of the more important differences
that need to be borne in mind.

1.  Behavioral differences, including the shell closure
response to high concentrations of pollutants in the water
(Davenport 1977).
2.  Differences in seasonal reproductive/nutrient
storage cycles.  For example, oysters generally synthesize
and store glycogen during the spring phytoplankton bloom (in
temperate waters), with spawning occurring in early summer;
mussels frequently lay up glycogen stores in the autumn and
use them for the production of gametes during the winter for
spawning in spring.  There are considerable temporal and
spatial variations in these patterns, however, and, as
discussed in another part of this chapter, we recommend that
attempts be made to elucidate these cycles in the different
areas of a monitoring program.

3. Morphological and cytological differences. At present, we are rather ignorant of the subtle histological differences between closely related species of bivalves; such differences exist, however, and can be fully evaluated only by experienced cytologists.

4. Biochemical differences. Among those of concern, it can be noted that species differ in their capacities for anaerobic metabolism during periods of stress (Widdows and others 1979a). There are differences also in patterns of nitrogen metabolism (Bayne and Scullard 1977) as well as in the seasonal cycling of glycogen and lipid stores.

## Tissue Differences

Knowledge of the distribution of contaminants within various tissues of the animal is a vital element in thorough assessment of the biological effects, but information to be gained from detailed analyses of this sort is often difficult to interpret. On the one hand, correlations between biological effect and the level of contaminant in the body may not be apparent with data from one tissue, but may be strikingly significant with data from a different tissue. On the other hand, differences in distribution between tissues may reflect differential detoxification capabilities that act either to remove the toxicant from a particular tissue (e.g., where metabolic transformation may modify the chemical nature of the xenobiotic) or to concentrate the contaminant in a certain tissue by sequestration in membrane-bound vesicles or by incorporation into inert proteins. The problems of detoxification are considered separately in this chapter, but we refer to them here to emphasize the complex nature of the relationship between tissue burdens and biological effects. In some cases, meaningful correlations can be found between pollutant concentrations in the entire organism and the biological effect (Widdows and others 1979a). In other circumstances, we suggest that knowledge of the tissue distribution will be essential for proper interpretation of the biological effects of the pollutant.

We recommend, therefore, that where a need for analysis of tissues is identified, the analyst should be encouraged to investigate separate tissues as well as the whole body.

## INDICES OF EFFECTS OF POLLUTION

The subject of the measurement of biological effects of pollutants on bivalves was recently considered by an ICES (1978) working party. The panel was in general agreement with the views expressed in the working party report and referred to it for much of the background on this topic. Changes induced in response to stress can be detected in animals from organismic to subcellular levels and may be morphological, physiological, or biochemical in nature. Some changes can be monitored with little time or effort and require no special equipment; others require special skills and sophisticated equipment. We will describe certain indices that seem particularly relevant to the measurement of stress in bivalves.

### Biochemical Indices

The biochemical indices of exposure to pollutants can be divided into two categories: (1) those that are indicative of the general stress response and (2) those that are indicative of the specific toxic action of particular pollutants. There is considerable information on the former; much less exists on the latter.

### General Indices of Stress

The general stress response to pollutant exposure involves the mobilization of energy reserves so that the mussel can both repair damaged tissues and synthesize substances used to detoxify pollutants. The three main categories of energy reserves used in mussels are carbohydrates, lipids, and proteins (Gabbott 1976). Therefore, a simple asessment of depletion of these energy reserves would appear to provide an index of pollutant stress. However, these biochemicals show seasonal variations in all types of organisms, and the variations are particularly high in mussels owing to the annual spawning cycle discussed earlier. Glycogen is available as an energy store prior to gametogenesis, but is depleted as the ripe gametes are produced. Protein is then the main energy store used during stress, until the carbohydrate and lipid reserves are built up again in the summer (Gabbott 1976). Despite the difficulties, Jeffries (1972) recorded substantial depletions of carbohydrates in clams from an estuary polluted with HCs, and Gabbott and Walker (1971) found that measurements of a simple condition index for oysters correlated well with measures of glycogen content. We suggest that, for routine monitoring purposes, measurement of condition index should suffice as a crude estimate of nutritional status. Where possible, these measurements should be accompanied by an assessment of reproductive stage to provide information on the seasonal cycle.

As a consequence of seasonal variability in availability of carbohydrate, lipid, and protein as energy reserves, there are seasonal variations in the ratio of oxygen consumed to ammonia nitrogen excreted (the O/N ratio) (Bayne 1976). During periods of protein synthesis (growth) the O/N ratio is high. In the winter and spring, when most carbohydrate and lipid energy reserves are channelled into the gametes, much more protein will be used for energy; then, the ratio of oxygen consumed to nitrogen excreted tends to be low (Bayne and Scullard 1977). Substantial decreases of the O/N ratio from the usual seasonal level may be a useful indicator of depletion of the normal carbohydrate and lipid energy reserves and are thereby indicative of stress.

Jeffries (1972) has proposed an index of stress in clams from HC polluted waters based upon the molar ratio of taurine to glycine. When this ratio was less than 3, the population was considered to be normal; when between 3 and 5, a chronic stress was indicated; and when greater than 5, the stress was considered to be acute. This simple index of stress has been confirmed by Bayne and others (1978). The theoretical basis for this increase has yet to be established, but the following comments are relevant: (a) with increasing stress, taurine concentration increases and glycine decreases; (b) taurine is used in the production of mucopolysaccharides, and glycine can be readily catabolized for the production of energy. D. Livingstone (National Environmental Research Council, Plymouth, U.K., personal communication, 1978) has found that the ratio of serine to threonine is responsive to stress and is low during stress.

Measures of this kind give easily and rapidly measured indications of the degree of stress. The more difficult and variable measures of depletion of carbohydrate, lipid, and protein reserves provide an indication of the overall chronic stresses to the organism. Pollution-induced depletions of energy reserves may be particularly important in decreasing the ability of the organism to survive periods of normal environmental stresses. For instance, if the lipid reserves are chronically depleted in mussels, gametes produced will be lipid deficient and, therefore, less viable (Bayne and others 1975). The mussel could also be less able to survive periods of prolonged starvation or environmental extremes of temperature, salinity, and oxygen depletion (Gabbott 1976, Bayne 1976).

A more specific toxic action of pollutants is that of decreasing the latency of lysosomal enzymes (Sternlieb and Goldfischer 1976, Moore 1977, Moore and others 1978a, Lowe and Moore 1979, Kohli and others 1977). Lysosomal enzymes are normally bound to the lysosomal membranes and glycoproteins in the lysosome so that they are effectively inert. Furthermore, the lysosomal membrane is normally nonpermeable to both the passage of lysosomal enzymes out of the lysosome and the passage of substrate molecules into the lysosome. However, when the organism is stressed, the lysosomal membrane is damaged in ways that are not fully understood, allowing a freer passage of enzymes into the cytoplasm concurrent with increased passage of substrate into the lysosomes. The latency of the lysosomal enzymes is thereby reduced. Decreases of lysosomal latency in the digestive gland of mussels result in a release of hydrolytic enzymes into the cytoplasm of the cell (Moore and others 1978b). Severe stress can result in releases of such high levels of lysosomal enzymes that there may be disruption of normal cellular constituents, with resulting autolysis and cell death (Amlacher 1970). Thus lysosomal destabilization gives a measure of the general response to stress that has consequences for the cellular condition of individual mussels and that is based upon the generalized mechanism of cytotoxicity of pollutants.

The measurement of lysosomal latency is based upon a cytochemical staining procedure that has currently been refined to a level where it is readily useable as a routine monitoring procedure (Moore 1976, Lowe and Moore 1979). Lysosomal latency is readily quantifiable and increases with increases in stress level (Moore and others 1978a). This index is a component of the general stress response in mussels, since lysosomal enzyme latency is responsive to a variety of environmental stressors, both natural and anthropogenic (Kohli and others 1977; Moore 1977; Bayne and others 1978; Moore and others 1978a,b; Moore 1976). It has been shown (Bayne and others 1979b) to correlate with physiological stress indices such as the scope for growth, and with the taurine/glycine ratio. The latency of lysosomal enzymes is subject to control by steroid hormones such as estradiol and progesterone (Moore and others 1978b), so some seasonal variability can be expected. It is nevertheless a powerful technique for monitoring biological effects and we recommend its inculsion in mussel watch programs.

Ivanovici (1978) has recently described the use of a biochemical measure, the adenylate energy charge (EC), that has been reported in a large variety of organisms under diverse environmental conditions but whose potential in impact assessment has been largely overlooked. The EC is calculated (Atkinson 1977) from the measured concentrations of the adenine nucleotides:

$$EC = (ATP + 1/2\ ADP)/(ATP + ADP + AMP)$$

where:

ATP = adenosine triphosphate,
ADP = adenosine diphosphate, and
AMP = adenosine monophosphate.

The index is a component of the general stress response, since it provides a measure of the metabolic energy that is potentially available to the animal at the time of sampling. Values range between 0 and 1. Values close to 1 indicate high metabolic potential, while lower values reflect decreased metabolic potential. Ivanovici and Wiebe (1978) point out four major advantages for the use of this index in assessing stress effects: (1) the adenine nucleotides are universally distributed in organisms and play a central role in energy metabolism; (2) variation between individuals is smaller for EC than for some other measures of metabolic intermediary substrates, allowing fewer replicates in the sampling program; (3) EC correlates well with physiological condition and potential for growth; and (4) the response time of EC to environmental change is fast, occurring in 24 hours or less in molluscs (Ivanovici 1978). In a review of the literature, Ivanovici and Wiebe (1978) suggest that EC values of between 0.8 and 1.0 are typical of healthy organisms. Values between 0.7 and 0.5 imply a stress condition although, at this level of stress, most organisms can still recover following an improvement in the environment. EC values of about 0.5 or less represent severe stress and considerably reduced possibilities of recovery.

Estimation of EC represents a potentially useful means of measuring stress and is a convenient characteristic of the general stress response. We suggest that further research is needed to catalogue the various conditions under which EC changes as a result of pollution in bivalves, but such research should, for most effective results, be integrated with the mussel watch programs.

Specific Toxic Effects

While much has been accomplished in identifying the general effects of pollutants on the stress response in organisms, less work has been done on the specific toxic actions of different pollutants. As we have pointed out earlier, assessment of the stress response in an area contaminated by many pollutants does little to tell us which pollutants are most detrimental to mussel health. An assessment of the effects of different pollutants at their biochemical sites of toxic action should enable us to elucidate more clearly which pollutants are exerting toxic effects. The following is a discussion of some known biochemical sites of toxic action for pollutants.

Metals One of the main sites of toxic action of metals appears to be enzyme function, particularly that of the metalloenzymes (Bremner 1974, Friedberg 1974). Toxic metals such as cadmium or mercury exert toxic effects by displacing essential metals such as copper and zinc from the metalloenzymes, resulting in conformational changes that prevent substrate molecules from fitting binding sites, thus rendering the enzyme inactive (Friedberg 1974). Alternatively, the catalytic subunit can be separated from the regulatory subunit of an enzyme so that the enzyme is no longer subject to feedback control (White and others 1968). It has been demonstrated that silver and mercury can deactivate digestive enzymes in vitro in _Mytilus galloprovincialis_ (Iordachescu and others 1978), but in no instances have enzyme activities been demonstrated to have been decreased in vivo in mussels due to metal exposure. Such effects have, however, been studied _in vivo_ in metal-exposed fish, where it was found that cadmium exposure resulted in increased synthesis of less effective zinc-containing enzymes (Gould 1977). This compensation effect has implications for increased utilization of energy reserves, a component of the general stress response. It should be noted that the enzyme changes occurred before there was any evidence of histopathological changes (Gould 1977). The results indicate that specific biochemical markers such as enzyme activity can be used to assess the

toxic effects of metals at their site of toxic action in
vivo. Further, detrimental changes at these sites of toxic
action act as a first indicator of toxic effects, before
gross changes are visible.

Disruption of ionic regulation has been noted in fish
exposed to mercurials (Schmidt-Nielsen and others 1977).
This is thought to occur via inhibition of the enzyme $Na^+$,
$K^+$-activated ATPase, an enzyme implicated in sodium
transport and ionic balance in a number of tissues. The
effects of metals on osmoregulation in mussels have not yet
been elucidated.

Metals such as cadmium and mercury are known to displace
$Ca^{++}$ from nerve membranes, resulting in leakage of $Na^+$ and
$H^+$ across the membranes; the result is that the nerve fiber
becomes closer to the threshold of action potential, i.e.,
requires less stimulus to elicit a response (Goldman 1970).
This could have important consequences in Mytilus edulis
where both stimulatory and inhibitory neurohormones may be
released from nerves. The branchial nerve has been
implicated in the control of ciliary activity of the bivalve
gill (Bayne 1976). Copper response has been shown to cause
respiratory and cardiovascular depression in M. edulis
(Scott and Major 1972). Brown and Newell (1972) have
suggested that this suppression is due to the inhibition of
ciliary activity rather than interference with respiratory
enzyme systems.

Petroleum Hydrocarbons  Petroleum HCs are thought to be
relatively nontoxic to mussels (Roberts 1976) although it
has been suggested in some instances that they can inhibit
filtering activity, elevate respiratory rates, and decrease
net carbon balance (Roberts 1976, Anderson and others 1974).
The main toxic action of petroleum HCs has been suggested to
involve release of hydrolytic enzymes from disrupted
lysosomes (Moore and others 1978a). Recent evidence
suggests that this effect is apparent at very low levels of
oil, between 10 and 40 $\mu g/l$ (Moore and others, unpublished
data, 1979).

There is some recent evidence to suggest that marine
organisms may be less sensitive to the toxic effects, and
more responsive to the carcinogenic effects, of chemical
pollutants such as petroleum (Stewart 1977). We suggest
that this may be the case for mussels also. Tumors have
been found in up to 69 percent of clams collected in areas
recovering from oil pollution (Brown and others 1977b) and
in Mytilus edulis in areas contaminated by carcinogenic
aromatic HCs (Lowe and Moore 1978). Yevich and Barszcz

(1976, 1977) have also found tumors in shellfish from oil-
polluted areas, and recently have noted dramatic increases
in the numbers of tumors in shellfish that have been exposed
in areas polluted in the past with oil but now considered to
be basically clean.

In view of the suspected involvement of lysosomal
enzymes in carcinogenesis (Allison 1969), it is suggested
that lysosomes can be considered to be a primary target for
petroleum HC toxicity. Because released lysosomal enzymes
are thought to cause cancer via disruption of chromosomes
(Allison and Paton 1965), we suggest that it is important to
monitor both latency of lysosomal hydrolases and chromosome
damage as sites of toxic action in oil-polluted areas.

Halogenated Hydrocarbons There appears to be no
information available on the specific site of toxic action
of halogenated HCs in mussels, although gross sites have
been distinguished. Laboratory exposures have resulted in
reduction of body weight, decreases of shell growth, reduced
feeding efficiency, reduced byssus formation, detrimental
effects on embryonic and larval development and survival,
elevated respiratory rates, and decreased condition index
(Roberts 1976). In Mercenaria mercenaria, metabolic
alterations have been found indicative of increased glucose
use and decreased gluconeogenesis (Roberts 1976). In
saltwater fish, DDT and PCBs disrupt normal osmoregulation
by inhibition of $Na^+$, $K^+$ and $Mg^{++}$ ATPases, resulting in
elevated $Na^+$ and $K^+$ levels (Anderson and others 1974). In
other animal studies, it has been found that toxic effects
of halogenated HCs occur when they bind to tissue
macromolecules (Brodie and others 1971, Shimada 1976).
Further, it appears that, as with petroleum HCs, halogenated
HCs can have a disruptive effect on lysosomes (Kohli and
others 1977). Research should be undertaken to elucidate
specific sites of toxic action in Mytilus edulis in order
that more effective measures of effects can be made.

Radionuclides Little information is available on the
effects of ionizing radiation on sites of toxic action in
invertebrates (Mix 1976). However, it can be suggested
that, as in other organisms, radiation can have a general
destructive action on cellular macromolecules and organelle
membranes. Specifically, it has been determined that
radiation results in releases of lysosomal hydrolytic
enzymes into the cytoplasm with resultant cytotoxic effects
(Allison 1969). Radiation also results in dramatic
increases in chromosomal breaks and other aberrations,

either as a direct result of the effects of radiation on chromosomes, or indirectly via the effect of released lysosomal enzymes (Allison and Paton 1965).

## Physiological Indices

Some of the components of the general stress response are physiological and represent changes in the main physiological attributes of the animal, such as its rates of feeding, respiration, and excretion, and the efficiency with which it absorbs nutrients from food. Indeed, the potential for growth, which is an integration of all these physiological changes, is the underlying characteristic of the stress syndrome. It is measured as the scope for growth.

### The Scope for Growth

This value is derived from measurements of rates of feeding, rates of oxygen consumption, and absorption efficiency. Integration is according to the balanced energy equation of Winberg (1960):

$$C = P + R + U + F$$

where:

C is total consumption of food energy,
P is production,
R is respiratory heat loss,
U is energy lost as excreta, and
F is fecal energy loss.

The absorbed ration, A, is the product of consumption, C, and the efficiency of absorption of energy from the food, e. Production may then be expressed as

$$P = A - (R + U).$$

When production, P, is estimated in this way it may be referred to as the scope for growth and the production of gametes. The use of this index in the assessment of the health or condition of mussels has been described by Bayne (1975) and Bayne and Widdows (1978). For example, Bayne and others (1979b) recorded the scope for growth in three

populations of mussels in 1976 and 1977. Individuals in the
population differed in their intensity of metabolism (rates
of oxygen consumption and excretion) and in both feeding
rates and absorption efficiency. When calculated as the
scope for growth and integrated over 1 year, the populations
differed as follows:

Swale:        32 kJ yr⁻¹,
Lynher:        8 kJ yr⁻¹, and
Kings Dock: 3 kJ yr⁻¹.

Bayne and others (1979b) also recorded a decline in ·the
scope for growth in one population that was measured
regularly for 5 years.

An example of the calculation of the scope for growth
from empirical data on feeding and respiration is given in
Appendix 6-A. The various components of the calculation can
be measured under field conditions when facilities permit.
For example, the study reported by Bayne and Widdows (1978)
involved physiological measurements made at sea from a
research ship. The previously discussed study of Bayne and
others (1979b) was carried out with physiological monitoring
equipment mounted in the back of a converted van used as a
mobile laboratory and driven to within a few meters of the
natural population on the shore (also see Bayne and Widdows
1978). In these cases, animals were sampled from a natural
population and used for physiological measurement under
conditions that very closely resemble those experienced in
the environment (similar or identical temperature, salinity,
oxygen and suspended particulate load). A wide size range
of animals is chosen as representative of the population and
analysis of the data is by linear regression on
logarithmically transformed values followed by covariance
analysis (Appendix 6-A). Bayne and Widdows (1978) have
discussed some of the variance associated with this
procedure. When measured under these circumstances, the
scope for growth provides a good prediction of actual
growth, as discussed by Bayne and others (1979b) and by B.L.
Bayne and C. Worrall (National Environmental Research
Council, Plymouth, U.K., personal communication, 1979).

However, where the objective is less to provide a good
prediction of growth and more to afford a comparison of
growth potential between different sites, a different
protocol may be adopted. In this protocol, mussels are
returned to the laboratory from the field and measurements
of feeding and respiration rates are made under standard
conditions in the laboratory seawater system. This protocol

has recently been evaluated by J. Widdows and P. Salkeld (National Environmental Research Council, Plymouth, U.K., personal communication, 1979). Three populations were visited and physiological determinations made following the normal practice of such field site visits. Individuals were then taken to the laboratory, where rate of oxygen consumption and rate of feeding were measured at a temperature and salinity approximating the field conditions at the time and at a suspended particle concentration of cultured algal cells (<u>Isochrysis</u> and <u>Phaeodactylum</u>) of $2 \times 10^5$ cells ml$^{-1}$. Agreement between the field and laboratory assessments was good.

This approach to physiological effects measurement is logistically much simpler than the procedure requiring measurements in the field; current results are encouraging and suggest that mussels maintain rates of key physiological processes in the laboratory for 2 to 3 days that are similar to rates in their natural populations. This finding opens the possibility of transplanting mussels for various periods of time along certain known or suspected pollution gradients, followed by transfer of the mussels back to the laboratory for physiological assessment. This approach was adopted by Widdows and others (1979b) in a pilot study. They transplanted mussels from a population at Conanicut Island, in Narragansett Bay, to several sites within the Bay, along a known pollution gradient (Phelps and others, U.S. EPA Laboratory, Narragansett, personal communication, 1978). After a period of 30 to 40 days, animals were returned to the laboratory, where physiological measurements of the scope for growth were carried out. The results of this study demonstrate a good correlation between degrees of contamination in Narragansett Bay and the scope for growth.

We consider that the scope for growth provides a powerful means for the measurement of the physiological condition of mussels, which can be employed in a variety of ways (field measurements, transfer of mature mussels to the laboratory for measurement, use of transplanted mussels) and involves simple physiological determinations. We recommend that, wherever possible, attempts be made to make these measurements the basic index of the stress response.


The Oxygen/Nitrogen Ratio

A characteristic property of the general stress response is the utilization of certain nutrient reserves to meet the enhanced energy demands resulting from stress. This response has certain biochemical consequences that are discussed elsewhere (e.g., depletion of carbohydrate and lipid stores, reduced energy charge). In addition, alterations in the balance between the catabolism of carbohydrate, protein, and lipid substrates may be discerned in an altered ratio between oxygen consumed and nitrogen excreted, when this O/N ratio is calculated in atomic equivalents (Bayne 1975).

Measurement of the O/N ratio is based on data for the rate of oxygen consumption (a component measurement in the scope for growth calculation) and the rate of ammonia excretion (Bayne and Widdows 1978). Low values (less than 10) signify a predominance of protein catabolism and represent a stressed condition; high values (in excess of 50) imply the catabolism of lipid and carbohydrate with a low proportion of protein breakdown, and this is indicative of a more healthy animal. There is a considerable seasonal variation in this index coincident with the seasonal cycles of reproduction and nutrient storage. However, at certain times of the year, particularly prior to spawning, differences in the O/N ratio may be informative and have been shown (Widdows and others 1979b) to correlate well, in transplanted mussels, with the pollution gradient in Narragansett Bay.

Fecundity and Egg Viability

We have suggested that for a physiological, or any other, change to be recognized as the result of stress, the animal must be shown to be disadvantaged in some fundamental way, such as in its fecundity. Fecundity and the chances of survival of the gametes are fundamental attributes of fitness and any effect on these factors will have consequences for the survival of the populations.

A good correlation between the scope for growth and fecundity has been demonstrated in laboratory experiments (Bayne and others 1978). In these studies not only was the number of eggs per female reduced at a low scope for growth, but the energy content of the eggs also declined as a result of a reduction in the amount of lipid packaged into each egg (Bayne and others 1978). Fecundity estimates are not easy to make for bivalves; in these studies individual animals were induced to spawn in the laboratory and the numbers of eggs counted. An alternative approach is to weigh the gonad (mantle tissue) and the whole body at different periods of the year (see the following section, "Condition Indices"),

since spawning is often detectable as a decline in weight of
the mantle, and this can be converted to an estimate of
fecundity. There are many potential errors here, however,
mostly due to incomplete spawning and to continued storage
of reserves even during periods of spawning. (This applies
particularly to species in warmer waters.) We do not
recommend fecundity estimates for routine monitoring, but
more research effort should be directed toward the
correlation of some stress indices and the fecundity of
individuals.

## Condition Indices

A useful measure of the physiological condition of
bivalve molluscs is their nutrient state. As discussed
earlier, stores of glycogen, lipid, and protein may be
mobilized to meet metabolic demands during stress.
Alterations in the concentrations of organic constituents
result from changes in availability of food, in rates of
feeding, and in rates of catabolism. Availability of food
is related to seasonal cycles in phytoplankton and detritus
production. Metabolic processes that require energy and may
reduce nutrient stores include respiration, feeding,
excretion, osmoregulation, cardiovascular function,
detoxification, and reproduction. Of these, reproductive
demands result in the greatest changes in concentration and
distribution of storage products.

Changes in nutrient stage have been quantified using the
condition index. Condition index is defined either as the
ratio of the dry-flesh weight to the total weight (flesh +
shell), or as the ratio of the dry-flesh weight to the
internal volume of the shell. Use of the former is
appropriate only when the relationship between dry flesh and
shell weight has been established. Using this simple index,
it has been possible to demonstrate for mussels and oysters
seasonal changes in nutrient state of both field populations
(Gabbott and Walker 1971) and of hatchery and laboratory
populations (Walne 1970, Bayne and Thompson 1970). The
pattern of change in condition index that generally occurs
is a rapid decrease upon spawning, a period of little
change, and then an increase to the pre-spawn condition.

The impact of a pulp mill on an oyster population in
Canada was assessed by monitoring the condition index (Ellis
1970, Anderson 1975). Data are available before and after
the discharge from the mill. The condition index in oysters
since 1955 has been determined in the spring (April or May)

of each year. A series of oysters was sampled from a
distance within 1 km of the outfall to 20 km. The condition
index of oysters located up to 2 km from the outfall
declined from pre-discharge levels of approximately 100 to a
level of 60 in 1970, which was 12 years after discharge
started in 1958. Over the same period, controls had a
condition index of approximately 100, but index reductions
(not tested statistically) in oysters located up to 7 km
from the outfall were suggested. In 1971, zinc
concentrations in oysters were monitored because a zinc-
based bleach (zinc hydrosulfite) was used by mills in the
area. A general relationship between high zinc levels in
oysters (up to 20 ppm dry weight) and low condition factor
(down to 70) was obvious. The zinc-based bleach was
withdrawn in 1973 by all mills in the area. After
withdrawal of the bleach at the mill that was monitored,
negative correlations between zinc concentrations in oysters
(to 20 ppm) and condition factors (to 40) were found from
data in 1974 and 1975. The slope of the curve that
describes this relationship became increasingly negative
($-4.65$ in 1974 to $-12.08$ in 1975) as zinc levels fell and
condition factors improved.

Another technique that has been used to assess nutrient
state is to determine ash-free, dry-flesh weight (dry weight
minus ash weight). Gabbott and Walker (1971) found that the
changes in this parameter were similar to those in condition
index in the adult oyster Ostrea edulis maintained under
hatchery conditions. Concentrations of zinc and the
condition index of oysters were found to be a function of
the discharge of materials from a pulp mill (Anderson 1975).
Data on these two parameters over a 5-year period were
correlated.

Changes in biochemical composition have also been
measured to determine the nutrient state of mussels and
oysters. Much of the work on seasonal changes in metabolic
stores has been reviewed by Gabbott (1976). Changes in the
total quantities of glycogen, lipid, and protein were found
to be similar to those of condition index. However, changes
in concentrations and distribution of the three storage
materials were controlled primarily by the demands of
gametogenesis.

Changes in nutrient state are reflected also in the
relative size of various tissues in the organism. Relative
size can be expressed as a body component index (BCI):

$$BCI = \frac{\text{dry weight of a tissue}}{\text{total dry weight of flesh}} .$$

Of particular importance in mussels and oysters is the gonad
index, which varies greatly in size during the reproductive
cycle. In mussels, gonadal tissue is distributed throughout
most of the body and is difficult to isolate completely from
other tissues. However, a significant fraction is present
in the mantle. Monitoring changes in gonad BCI should
provide an index that reflects changes in reproductive
state.

Observations of changes in gonadal index in bivalve
molluscs maintained under nutritive or temperature stress in
hatcheries or in laboratories indicate that alterations do
occur in the typical reproductive cycle. Depending on the
stage of the reproductive cycle and the magnitude of the
stress, changes ensue that reduce the reproductive capacity
of the organisms to different degrees. Phenomena that have
been observed are the absence of spawning with a gradual
recession of the gonadal mass, reduction of the total
quantities of gametes released, and failure of development
in the larvae from stressed adults (Bayne and others 1978).

Indices Based on Histopathology

The techniques of histopathology have enormous potential
in monitoring the health and condition of animals, and have
been employed to this end for many years in medical and
veterinary fields. Routine application of these techniques
to bivalves suffers from one major disadvantage, namely our
ignorance of the normal morphology and histology of many
species. Consequently, judgments about the extent of
pathological damage to cells and tissues can only be made
with confidence following some years of experience in
looking at molluscan material. This limits the ubiquity
with which histopathology can be adopted in monitoring
programs. Nevertheless, there is considerable research
extant on the cytology and pathology of stress in bivalves,
including studies on mussel watch programs in the United
States and in Great Britain, so information is being
accumulated rapidly.

Eight categories of pathological condition in oysters
and mussels have been studied in recent years and are
discussed below. In addition, Appendix 6-B includes a brief
description of techniques for the preparation of molluscan

tissue for histopathological examination, and Appendix 6-C presents the results of an "inter-collaboration" exercise in the assessment of the pathological condition of samples taken from the U.S. Mussel Watch Program (Yevich and Barszcz 1976). In this exercise, the same slides--of material from 10 populations of Mytilus--were examined independently by two laboratories, and the results ranked according to mussel condition. The two laboratories used somewhat different criteria in making their assessments, but the degree of agreement between the two offers hope for the general application of these techniques in monitoring. The panel feels very strongly that further resources should be allocated to histological examination of invertebrate material with a view to adding to our information on both the normal and stressed condition.

Our brief account of eight histopathological conditions follows.

1. Hyaline degeneration of the connective tissue of the gills. This condition has been observed in animals exposed to oil both in the laboratory and at the site of an oil spill. It is illustrated in Figure 6.1.

2. Parasite burden. Extensive parasitic infestation may indicate a stressful condition in the host, not necessarily because the parasites themselves impose a stress, but because bivalves under chronic stress from the environment, including pollution, may be less able to resist parasitic infestation than healthier animals. Barszcz and others (1978) noted an increase in the number of parasites in oysters exposed to various types of crude oil, and studies on other phyla have also indicated a relationship between exposure to a pollutant and parasite burden (e.g., Vernberg and Vernberg 1963, Boyce and Yamada 1977).

In this context, the parasite burden in the individual is considered extensive when the animal contains three to five different species of parasite, incuding larval trematodes, nematodes, ciliates, microsporidia, and viruses. In many cases, a parasitic infection does not induce an inflammatory response by the host (Moore and Lowe 1977, Lowe and Moore 1978), and under these circumstances no damage to the host can be claimed. In other cases a severe host response is apparent as an inflammation of the connective tissue and invasion of the area by amebocytic hemocytes. In a few examples, parasitic infestation, assessed both as numbers of parasites and as the severity of the host response, may be useful in assessing bivalve health.

187

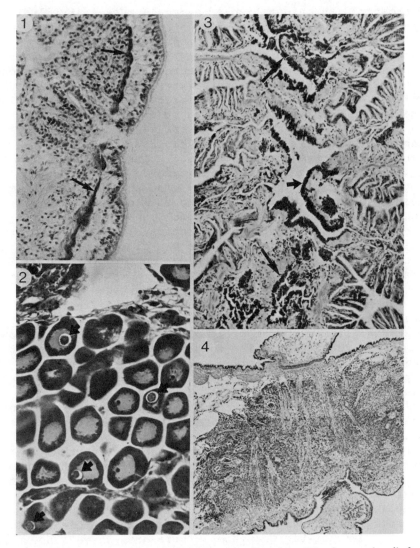

FIGURE 6.1 Oyster (*Crassostrea virginica*) gill exposed to Nigerian crude oil for 9 months at a concentration of 4 ppm (Yevich 1976). NOTE: Arrows point to hyaline degeneration at the connective tissue under the epithelium of the food groove of the gills.

FIGURE 6.2 Reproductive tract of female mussel (*Mytilus edulis*) (Yevich 1976). Arrows point to ova containing microsporidia.

FIGURE 6.3 Gill of oyster (*Ostrea edulis*) collected from the Amoco Cadiz oil spill site, 1 month after spill occurred (Yevich 1976). NOTE: Arrows point to wall of water tubes which are lined with mucus-secreting cells.

FIGURE 6.4 Soft shell clam (*Mya arenaria*) collected from Seaport, Maine, U.S.A. (Yevich 1976). NOTE: Over 33 percent of the body has been invaded by neoplastic cells.

Microsporidian infestation of a female mussel (<u>Mytilus</u> <u>edulis</u>) is shown in Figure 6.2.

3. <u>Stimulation of mucous production by the gill</u>. Laboratory and field studies show that molluscs and fish exposed to oil demonstrate an increase in mucous secretory activity and that the mucous ciliary mechanisms may act as a defense against oil. Several different mucopolysaccharides are produced by molluscs exposed to crude oil and there is an increase in the number of mucous secretory cells. Laboratory studies conducted by Manfredi (U.S. EPA Laboratory, Narragansett, personal communication, 1978) on soft-shell clams, <u>Mya arenaria</u>, exposed to varying concentrations of No. 2 fuel oil also showed an increase in mucous secretory cells containing several different mucopolysaccharides. Stainken (1975) has shown that mucosubstances bind with oil micelles, forming a complex that is ejected from the animals via ciliary pathways. An example of increased mucous production in bivalve gill tissue is shown in Figure 6.3.

4. <u>Gonadal neoplasm</u>. Barry and Yevich (1975) and Yevich and Barszcz (1977) have shown a 1 to 22 percent incidence of gonadal tumors in soft-shell clams (<u>Mya</u> <u>arenaria</u>) from the site of an oil spill off the Maine coast in the United States. The tumor, found in both males and females, involved the germinal epithelium of the reproductive tract. While some of the animals had only small foci of neoplastic cells in the reproductive follicles, in the majority of the clams the follicles were filled with neoplastic cells, causing distension and loss of distinguishing sexual characteristics. Metastasis of the neoplastic cells to other areas of the body such as the gills and kidney was seen in over 40 percent of the animals with tumors. In one animal, the metastasis was so extensive that over one-third of the body was taken over by the neoplastic cells (Figure 6.4). The only distinguishing tissues in the area of metastasis were the muscle bundles.

5. <u>Hemopoietic neoplasm</u>. There are, as yet, no established criteria with which invertebrate neoplasia can be diagnosed. However, Pauley (1969), in his review of molluscan neoplasia, suggested that the infiltration, invasion, or replacement of normal cells by actively mitotic, atypical cells was indicative of a potential neoplastic condition. Suspect neoplasms in marine invertebrates that fulfill Pauley's criteria have been reported in a number of bivalves, including <u>Ostrea</u> <u>lurida</u> (Mix 1975, 1976), <u>Crassostrea</u> <u>virginica</u> and <u>C</u>. <u>gigas</u>, and <u>Mytilus</u> <u>edulis</u> (Lowe and Moore 1978). The condition in

mussels was characterized by the infiltration and replacement of the vesicular connective tissues by enlarged atypical, mitotically active, basophilic hemocyte-like, periodic-acid-Schiff-negative cells (Figure 6.5). These abnormal cells, which were rich in cytoplasmic RNA and have significantly higher DNA values than normal hemocytes, were observed to cause degeneration of the digestive cells. The replacement of the vesicular connective tissue would ultimately lead to a loss of glycogen storage capacity with resulting cessation or loss of gametogenic capability. Similarly, the degeneration of the digestive cells would gradually lead to a loss of intracellular digestive capability and ultimately probable starvation and death.

6. Granulocytomas. Granulocytomas, so called because of their dominant cell type, are non-neoplastic inflammatory responses to a range of environmental pollutants. The condition, first described by Lowe and Moore (1979), was observed in the digestive gland and mantle tissue of Mytilus edulis and differentiated itself from hemocyte responses to invading tissue parasites in that while the latter are composed of macrophages and lymphocytes, the granulocytomas are composed of granulocytes that exhibit coagulation of the cytoplasmic granules (Figure 6.6). The condition originates in the hemolymph ducts, but expands--overcoming attempts at encapsulation--and invades the surrounding connective tissues, inducing atrophy and autolysis of the digestive tubule epithelial cells. In a survey of eight populations of mussels there appears to be a good correlation between the incidence of granulocytomas and the levels of anthropogenic stress (unpublished data). In view of the degenerative effects of the digestive cells, this condition is thought to be indicative of a general loss in condition.

7. Hemocytic infiltration of tissues. Localized increases in all, or specific, hemocytes can be induced by a variety of factors including parasitism, injury such as shell damage (Des Voigne and Sparks 1968, Bubel and others 1978, Bayne and others 1979a), or even the natural process of tissue resorption following spawning (Bayne and others 1978). However, whether these localized increases involve the production of additional blood cells or represent a redeployment of the existing blood cell complement is not always apparent. Increases in hemocyte numbers in the response to elevated temperature were demonstrated in Crassostrea virginica by Feng (1965), who postulated that as hemocytes participate in the processes of digestion their numbers may fluctuate depending on whether or not the animals are feeding. Thompson and others (1978), working

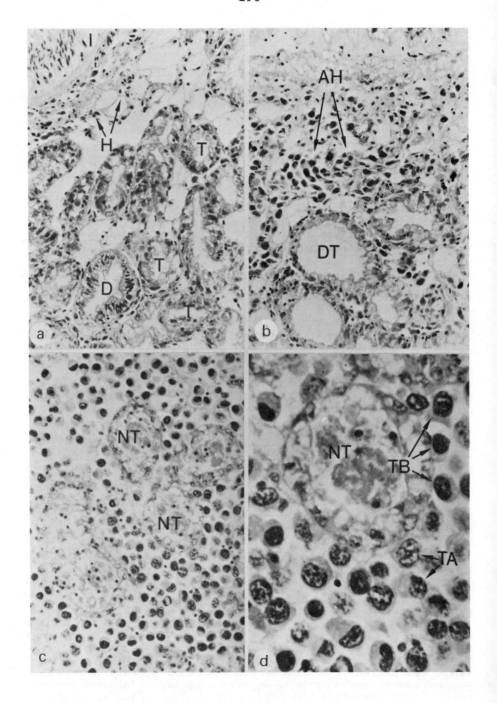

with <u>Mytilus californianus</u>, showed that the change from starvation to high food ration did not induce an increase in hemocyte numbers, and neither did a combination of increased ration and increased temperature. However, exposure to air at 21°C led to an immediate increase that returned to normal once the animals were returned to water. Clearly, the number of hemocytes in the animal may vary at any time depending on natural factors such as tissue resorption, feeding rhythms, and parasitism; however, these hemocytic responses are usually focal and of a transient nature. While this is the case and hemocyte numbers generally remain within acceptable limits, bearing in mind natural variability, this is not viewed as being indicative of a loss of condition. Once the hemocyte numbers exceed acceptable limits and the cells are evident in very large numbers throughout the whole connective tissue network, this is taken to be a manifestation of stress and is classified as being indicative of loss of condition.

    8. <u>Loss of synchrony in digestive tubules</u>. The principal phases in the digestive cells of molluscs have been characterized as adsorption, digestion, fragmentation, and excretion (Owen 1966, Langton 1975, Bayne and others 1976). While in some species these phases are synchronized so that at any time all digestive cells are in the same phase (Langton 1975), this is not so with all bivalves. Langton (1975) has shown that all the various phases of the digestive cells can be found at any time during the tidal cycle. This is obviously a situation more conductive for an animal that is feeding almost continuously.

    Digestion in bivalves has been shown to occur in two sites. Extracellular digestion is brought about by the action of the crystalline style in the stomach, and this is followed by intracellular digestion in the digestive cells in the digestive diverticula. In <u>Crassostrea virginica</u>, <u>C. gigas</u>, and <u>Ostrea edulis</u> it has been shown that the crystalline style dissolves and reforms every tidal cycle

---

FIGURE 6.5 Hemopoietic neoplasia in mussels characterized by infiltration and replacement of the vesicular connective tissues by enlarged atypical, mitotically active cells (Lowe and Moore 1979). *a,* A section through normal digestive gland tissues of *Mytilus edulis* showing the digestive tubules (T), ducts (D), and hemocytes (H) in the connective tissues underlying the intestine (I), magnification 3.2×. *b,* A section as above showing degenerating digestive tubules (DT) and infiltration of the connective tissue underlying the intestine by enlarged atypical (neoplastic) hemocytes (AH), magnification 312×. *c,* A section showing severe infiltration of the connective tissues by atypical (neoplastic) hemocytes with associated necrotic digestive tubules (NT), magnification 3.2×. *d,* A section showing a necrotic digestive tubule (NT) with infiltrating type A (TA) and type B (TB) atypical hemocytes, magnification 788×.

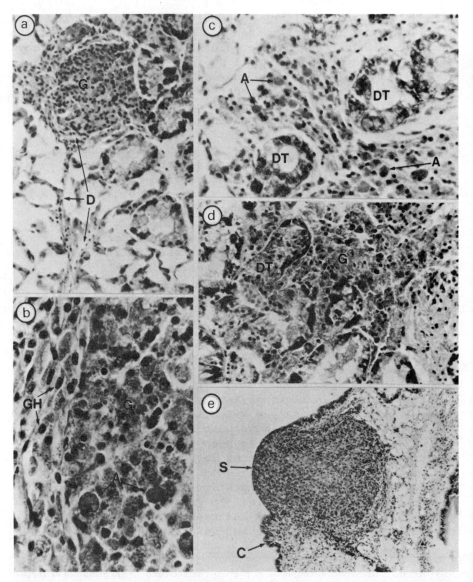

FIGURE 6.6  Granulocytomas in *Mytilus edulis* (Lowe and Moore 1979). *a,* Section of the digestive gland of *Mytilus edulis* showing a small granulocytoma (G) within a dilated hemolymph duct (D), magnification 320×. *b,* Section of a granulocytoma (G) showing amorphous bodies (A) and the surrounding layer of fusiform granulocytes (GH), magnification 800×. *c,* Section showing granulocytoma cells, with associated amorphous bodies (A), invading the digestive gland tissues between the digestive tubules (DT), magnification 500×. *d,* Section showing the breakdown of the digestive tubules, adjacent to invading granulocytoma cells (G), resulting in a loss of cellular integrity, magnification 320×. *e,* Section showing metaplastic changes in the epidermal cells of the digestive gland from ciliated columnar (C) to simple squamous (S), magnification 125×.

(Morton 1971, Bernhard 1973, Langton and Gabbott 1974).
However, it has also been shown in <u>Cardium</u> <u>edule</u> and <u>Mytilus</u>
<u>edulis</u> that phasing of the digestive cells is not linked to
the tidal cycle but to the influx of food (Owen 1972,
Langton 1975); of course, in the case of littoral animals
the presence of food is dependent on the tidal cycle (Morton
1977).

Several stressors have been shown to upset the digestive
rhythm in mussels; these include spawning, which induces
structural alterations in the digestive cells resulting in
autolytic changes (Bayne and others 1978), and starvation
(Thompson and others 1974), which results in altered
structure correlated with a decline in the digestive gland
index.  Elevated temperature has also been shown to alter
tubule structure in <u>Mytilus</u> <u>californianus</u> (Thompson and
others 1978), and Moore and others (1978 a,b) reported
structural alterations in the digestive cells of mussels in
response to injected anthracen and estradiol-17$\beta$.

Routine screening of sections of the digestive
diverticula of mussels from environments of different
qualities has indicated that all the tubules emanating from
one duct are in the same phase (Lowe, D., National
Environmental Research Council, Plymouth, U.K., personal
communication, 1978).  This observation is supported by the
findings of Langton (1975) that food is passed into small
groups of tubules served by the same duct, therefore
indicating that their digestive cells would be active while
those not served by that duct may be in a different phase.
Mussels from environments that are known to be highly
polluted have been observed to exhibit a greater number of
digestive cells in the fragmentation phase (Lowe, D.,
National Environmental Research Council, Plymouth, U.K.,
personal communication, 1978), and this observation is
supported by the work of Moore and others (1979) who found
that the lysosomal stability of the digestive cells
decreased with an increasing degree of stress.  In terms of
digestive cell morphology, this phenomenon manifests itself
as an increase in the numbers of digestive cells in the
fragmentation phase (Moore and others 1978a).

Experimental evidence has indicated that changes in the
digestive rhythm can be induced by stress with a resulting
loss of synchrony of digestive cell phasing (Figure 6.7).
Bayne and others (1976) showed a good correlation between
scope for growth and lysosomal stability.  As digestive cell
fragmentation and decreased lysosomal stability represent
aspects of the same phenomenon, it might also be expected
that an increase in the incidence of digestive cells

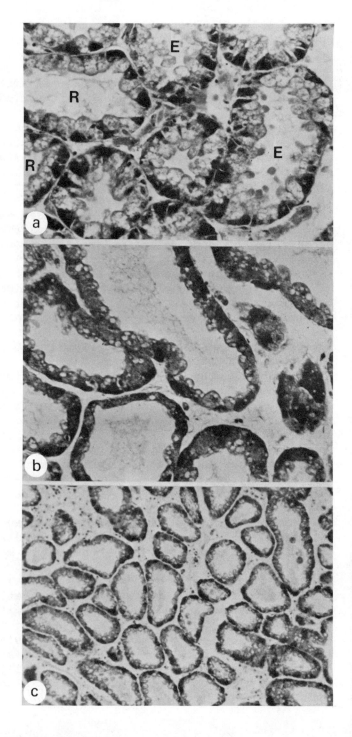

exhibiting fragmentation is indicative of reduced scope for growth and therefore of loss of condition.

## Genetic Indices

The breakage of chromosomes or chromatids by irradiation or by chemicals, and their recombination to form aberrations, is well known. However, the number of studies of genetic effects on aquatic organisms is small and the quantities of pollutants used were generally in excess of those found in most polluted environments. Blaylock and Trabalka (1978) have reviewed the literature on genetic effects of irradiation on aquatic organisms, and indicators of change in genetic material in sperm from abalone exposed to increased concentrations of copper in the water were reported by Lake and others (1977).

Recently, techniques have been developed to determine changes in genetic material in mammals exposed to pollutants. The techniques permit easier and better resolution of chromosome, chromatid, and DNA alterations. Chromosomes of bivalve molluscs are well suited for chromosomal analysis because of the size and number of the chromosomes involved (Wada 1978). Application of the technique developed for mammalian tissues should provide information to assess the genetic damage to oysters and mussels of increased levels of pollution in aquatic ecosystems.

## DETOXIFICATION MECHANISMS IN BIVALVES

The general stress response tells us of the combined effects of all pollutants on the organism. Specific biochemical indices provide a more specific evaluation of the effects of individual pollutants. However, because there is considerable overlap of the sites of toxic action

FIGURE 6.7 Histopathological demonstration of changes in molluscan digestive rhythm induced by stress (Lowe and Moore 1979). a, Digestive tubules in control mussels exhibiting the resting phase (R) and the excretory phase (E) of the digestive rhythm, magnification 320X. b, Digestive tubules in experimental mussels exhibiting a marked reduction in epithelial cell cytoplasm and a resulting loss of digestive synchrony, magnification 320X. c, Same tissue at low power showing the distribution of tubules throughout the digestive gland that exhibit a loss of digestive synchrony.

of various pollutants, further refinements may be necessary before the relative contribution of each pollutant to toxicity in various polluted areas can be assessed. We suggest that further study of this topic should involve an assessment of the partitioning of a pollutant between its detoxification system and its site of toxic action. Specifically, we recommend quantification of the levels of each pollutant in its detoxification system and at its site of toxic action. As will be described, studies have indicated that it is possible to obtain a measure of the loading capacity of certain detoxification systems. Since it is possible to measure the loading capacity of these systems, it should also be possible to estimate how much more pollutant a partially loaded detoxification system can accommodate before it becomes saturated. It is apparent with respect to at least two different detoxification systems (metallothionein and lysosomes) that toxic effects of pollutants do not occur until the loading capacity of the detoxification systems is surpassed (Brown and Parsons 1978, Moore 1977). The saturation level of the detoxification system corresponds to the "threshold level" so often described for various pollutants (Maugh 1978a,b; Brown and Parsons 1978). The following is a discussion of what is known about the detoxification systems available for each of the four categories of pollutants being measured in the mussel watch.

## Metals

### Metallothioneins

Trace metals are detoxified in organisms by binding to the protein metallothionein (Piscator 1964, Brown and others 1977a) or by sequestration by lysosomes (Sternlieb and Goldfischer 1976, Moore 1977). Copper and zinc are essential components of many metalloenzymes, and therefore will be found in the enzyme-containing pool (Brown and others 1977a, Brown and Chatel 1978a). Excesses of copper and zinc above those levels necessary for metalloenzyme functioning, and toxic trace metals such as mercury, cadmium, silver, and tin, are bound to metallothionein and thereby rendered biologically inert (Brown and Chatel 1978a, Winge and others 1975). Copper and zinc are necessary for the functioning of many enzymes, as they maintain both tertiary and quaternary protein structure. If metalloenzymes are exposed to an excess of the functional

metal or to competing trace metals such as mercury or
cadmium, they may lose their ability to function normally.
Dysfunction may be a result of conformational changes
resulting from interference of excess metals, or toxic
metals, with essential metals in the metalloenzyme (Bremner
1974, Friedberg 1974). Therefore, to ensure proper cellular
functioning, excesses of essential metals, or toxic metals,
must be removed from the zone of biological activity. This
is accomplished by production of metallothionein within the
organism.

Studies with zooplankton, fish, mice, and mussels
(Mytilus edulis) have indicated that levels of metal bound
to metallothionein will increase with increasing exposure
level until a plateau of metallothionein level is reached
(Table 6.1). At this point there will be a spillover of the
metal into the enzyme-containing pool (Brown and others
1977a, Brown and Parsons 1978). It appears as though at a
certain level of metal exposure, the binding capacity of
metallothionein will be surpassed. Only when the binding
capacity of metallothionein is surpassed, and toxic trace
metals occur in the enzyme-containing pool, do visible toxic
effects occur (e.g., decreased growth rate and tissue
necrosis). These effects are apparently due to the
interference of toxic metals with metalloenzymes (Chen and
Ganther 1975, Prohaska and others 1977).

Various reports have indicated that Mytilus edulis has a
substantial capacity to detoxify toxic trace metals via
metallothionein, particularly cadmium. Talbot and Magee
(1978) sampled Mytilus edulis from Corio Bay, Australia, an
area known to be polluted with high levels of cadmium, and
they indicate that the levels of cadmium in these mussels
are among the highest in the world. They showed that most
cadmium is bound to metallothionein, slightly lesser amounts
to the low-molecular-weight cytoplasmic pool, and least
amounts to the enzyme-containing pool. When Noel-Lambot
(1976) exposed Mytilus edulis to 0.005 ppm cadmium for 90
days in the laboratory, most cadmium was bound to
metallothionein. When exposed to 0.13 ppm cadmium for 36
days, very high levels of cadmium were bound to
metallothionein, with lesser but substantial levels in the
enzyme-containing pool. Noel-Lambot does not indicate if
these higher levels of cadmium in the enzyme-containing pool
correlate with toxic effects in the mussels.

From studies by Brown and Chatel (1978c), it is apparent
that survival of Mytilus edulis is reduced when much higher
than normal levels of trace metals occur in the enzyme-
containing pool (Table 6.1); this occurs when the binding

TABLE 6.1   Increases of Metal on Metallothionein and in the Enzyme-Containing Pool with Increasing Exposure Level

| Organism | Zooplankton[a] | Fish[a] | Mice[b] | Mytilus edulis[c] | Mytilus edulis[d] |
|---|---|---|---|---|---|
| Metallothionein-bound metal ($\mu$mole/g tissue, wet weight | Hg | Hg | Cd | Cd + Cu + Zn | Cd + Cu + Zn |
| Control | $0.04 \times 10^{-4}$ | $0.29 \times 10^{-3}$ | 0.0024 | 0.061 | 0.067 |
| Low exposure[e] | $0.27 \times 10^{-4}$ | $0.35 \times 10^{-3}$ | 0.0655 | 0.092 | 0.087 |
| High exposure | $0.26 \times 10^{-4}$ | $0.38 \times 10^{-3}$ | 0.0678 | 0.098 | 0.222 |
| Enzyme-containing pool ($\mu$mole/g tissue, wet weight) | | | | | |
| Control | $0.32 \times 10^{-4}$ | $0.11 \times 10^{-3}$ | 0.0004 | 0.295 | 0.244 |
| Low exposure | $0.40 \times 10^{-4}$ | $0.23 \times 10^{-3}$ | 0.0073 | 0.349 | 0.249 |
| High exposure | $1.17 \times 10^{-4 f}$ | $4.04 \times 10^{-3 g}$ | $0.0822^h$ | $0.614^i$ | $0.710^j$ |

[a]Exposed to trace, 1 and 5 $\mu$g Hg/L in a controlled ecosystem for 72 days (Brown and Parsons 1978). Each zooplankton sample was a composite of 2 g of whole zooplankton. Each fish sample was a composite of 3 fish livers.

[b]Exposed to trace, 50 and 200 mg Cd/L in their drinking water for 28 days (Brown and Chatel 1978b, *In* Brown 1978). Each sample was a composite of 3 mouse livers.

[c]Exposed to trace, 0.161 and 1.61 mg  CD+Cu+Zn /L in ratio Cd:Cu:Zn of 0.1:9:7 for 14 days in the laboratory (Brown and Chatel 1978c, *In* Brown 1978). Each sample was a composite of the soft parts of 15 mussels.

[d]Chronically exposed to a variety of metals in the vicinity of a sewer outfall where Cd, Cu, and Zn were in ratio of 0.1:9:7 in sediments at the high exposure station. Distances from the sewer outfall were: controls, 4.5 Km; low exposure, 2.5 Km; and high exposure, 1.5 Km (Brown and Chatel 1978c, *In* Brown 1978). Each sample was a composite of the soft parts of 15 mussels.

[e]There were no readily apparent toxic effects to low exposure animals.

[f]Zooplankton counts in high exposure were one-fourth those of the controls or low exposure on day 72.

[g]Fish weights in high exposure were one half those of the controls or low exposure on day 72.

[h]Liver weights were reduced by 11% in high-exposure mice relative to control or low exposure mice. Two of six high-exposed mice died during the 28-day exposure.

[i]High-exposed mussels did not survive the exposure period while there were no apparent mortalities in controls or low exposures.

[j]High exposure area was very sparsely populated by mussels while control and low exposure areas were richly populated.

capacity of metallothionein is apparently surpassed. Furthermore, it appears as though the capacity of Mytilus edulis to detoxify trace metals by production of metallothionein is much less with an acute exposure than with a chronic exposure, since after 14 days of high metal

exposure in the laboratory, metal levels bound to
metallothionein were increased 1.6-fold over control values;
while, after chronic metal exposure in the environment,
metal levels bound to metallothionein were increased 3.3-
fold over control values.

Thus, it appears that in mussels, as in other organisms,
the levels of metals on metallothionein increase with
increasing exposure level, and toxic effects do not occur
until the binding capacity of metallothionein is surpassed
and metals spill over into the enzyme-containing pool.
Since techniques are available to quantify the levels of
metal on metallothionein or in the enzyme-containing pool,
it is recommended that these procedures be implemented into
the current mussel watch program.


Lysosomes

Metals are also known to be detoxified by partitioning
into lysosomes in Mytilus edulis in kidney, digestive gland,
gut, gill, and blood cells (Lowe and Moore 1979).  Mussels
(Mytilus edulis) in a zinc-polluted environment were
demonstrated by cytochemical means to contain elevated
levels of zinc in lysosomes (Lowe and Moore 1979).  The
relationship between detoxification of metals in mussels by
sequestration by metallothionein or lysosomes is not clear
at present; however, it has been found in higher organisms
that metallothionein is stored in lysosomes (Porter 1974).
Lysosomal storage is possible for the protein
metallothionein since it is resistant to degradation by
proteases (Webb 1972).  Lowe and Moore (1979) have found
that the metal-containing lysosomes may serve as excretory
vesicles for metals from the kidneys of Mytilus edulis.

It has been suggested that when the storage capacity of
the lysosome for metals is surpassed, the metals will be
released into the cytoplasm (Moore 1977, Goldfischer 1965).
Furthermore, it has been found in hydroids and Mytilus
edulis that, as the levels of metals in lysosomes increase,
the lysosomal latency decreases, resulting in release of
hydrolytic enzymes into the cytoplasm (Moore 1977, Lowe and
Moore 1979).  In hydroids, lysosomal B-hexosaminidase
activity increases approximately two-fold with increasing
metal load (Cu, Cd, Hg) (Moore and Stebbing 1976).  At
present, it may not be clear whether these toxic effects are
due to released hydrolytic enzymes, released metals, or
both.  Further research is needed to establish whether
metallothionein or lysosomes or both are responsible for

metal detoxification, and the relationship of lysosomal
storage of metals to metallothionein binding of metals.  Of
particular importance to an understanding of the toxicology
of trace metals may be the role of lysosomes in excretion of
metals.  Since metals result in decreases of lysosomal
latency, cytochemical procedures for measuring these
parameters are essential for an assessment of the effects of
metals on mussel health.


Cellular Mechanisms

   It is well known that mussels, oysters, and other
bivalves can tolerate very high concentrations of some
metals in their tissues without any apparent signs of
toxicity.  For example, Thrower and Eustace (1973) have
reported copper and zinc concentrations as high as 450 and
21,000 µg/g wet weight, respectively, in oysters from the
Derwent River estuary.  Clearly these animals must possess
efficient means of detoxifying the metals, preventing
interaction between the metals and certain enzymes in the
cell.  Possible cellular detoxification mechanisms have been
studied extensively by George, Pirie, and Coombs and their
colleagues in Aberdeen, using techniques of radioactive
labelling and analytical electron microscopy.  For example,
in a study of the accumulation and excretion of ferric
hydroxide by Mytilus edulis, George and others (1976)
concluded that uptake was by pinocytosis in the gills, gut,
and kidney and incorporation into membrane-bound vesicles
that resemble secondary lysosomes.  The absorbed iron was
exocytosed from the epithelial cells and then passed to the
amoebocytes in the hemolymph, after which the iron was
further distributed around the body.  In a similar study on
copper and zinc in the oyster, Ostrea edulis, George and
others (1978) were able to discriminate between two types of
amebocytic hemocyte:  granular acidophilic cells accumulated
copper and granular basophilic cells accumulated zinc.  The
potential toxicity of the metals was reduced by their active
uptake from the serum into the hemocytes (where
concentrations could be as high as 1,400 to 13,000 ppm
copper and 9,900 to 25,000 ppm zinc, wet weight).  Once the
metals had been pinocytosed into the hemocytes they were
further detoxified by compartmentation within membrane-bound
vesicles (lysosomes), just as for iron in mussels.  These
results are similar to findings by Moore (1977) and Lowe and
Moore (1979), referred to above.

Evidence suggests that this basic detoxification mechanism of pinocytosis and membrane-bound compartmentation, either directly into the hemocytes or via epithelial cells in the gut and gills, is responsive to additional chemical insult. Ruddell and Rains (1975) have shown that the number of basophilic hemocytes found in the mantle of <u>Crassostrea</u> is directly proportional to the total zinc in the tissue. George and others (1978) conclude that the detoxification process includes increased production of hemocytes in response to increased metal exposure, as well as the production of soluble extracellular products capable of complexing the metals in the serum and further aiding their overall detoxification.

These metal-laden hemocytes may eventually transport the metal to the kidney for release through exocytosis into the urine (Lowe and Moore 1979). However, in theory at least, the capacity of the animal to produce more hemocytes, or the capacity of the cells to produce more lysosomal sequestration without gross disturbance to normal cellular function, will eventually be exceeded, at which time more direct toxic effects can be expected to occur. When we know the capacity of these detoxification systems we will be able to predict with more confidence the dosages at which toxic effects can be expected.

There are circumstances, however, under which toxic effects can be expected before the detoxification capacity is exceeded. Hemocytes are actively involved in many aspects of wound repair in bivalves (Ruddell 1971, Bubel and others 1978). One hemocyte response to wounding involves the discharge of cellular contents of the hemocytes at the site of tissue damage; this is a component of the inflammatory response of bivalves (Des Voigne and Sparks 1968). Ruddell (1971) has recorded that hemocytes participating in this response may discharge metals in addition to other, presumably more functional, cellular components. This may represent a breakdown of the detoxification system with resulting exacerbation of a tissue trauma. The result is apparent as tissue damage, possibly under circumstances that are not normally harmful to the animal, such as during certain types of parasitic infestation.

In addition, evidence is accumulating that some forms of metabolic detoxification are centered on the hemocytes in bivalves (Bayne and others 1979a). Any response that saturates these hemocytes with metals may render them less efficient at other processes of detoxification, leading to supra-additive (truly synergistic) effects of pollution.

These are areas that require further research that focuses on the various detoxification processes and their maximum capacity for coping with chemical insult.

## Petroleum and Halogenated Hydrocarbons

In most organisms, HCs are removed from potential sites of toxic action by partitioning them into lipid pools (Allen and others 1974, 1976; Whittle and others 1977). HCs are then transported with blood lipid to the liver or another equivalent organ where they are sequestered into lipid vacuoles (Allen and others 1974, 1975; Kimbrough and others 1972). Cytochrome P450-linked reactions then convert these hydrocarbons in arene oxides (Jerina and Daley 1974, Shimada 1976, Bend and others 1977). These arene oxides are then converted into trans-dihydrodiols by epoxide hydrases or into glutathione conjugates, either by glutathione-s-epoxide transferase or by spontaneous reaction with glutathione (Jerina and Daley 1974, Shimada 1976, Brodie and others 1971). It is apparent that the arene oxide intermediates are both the more toxic and carcinogenic forms of the HCs (Oesch and others 1976, Jerina and Daley 1974, Miller 1970, Shimada 1976, Brodie and others 1971). Thus epoxidation of HCs is referred to as bioactivation. Subsequent action of the epoxide hydrases and glutathione-s-epoxide transferases converts these bioactivated forms into deactivated forms (Oesch and others 1976, Jerina and Daley 1974, Bend and others 1977). Whereas the original accumulated HCs are hydrophobic and partitioned into lipid tissues and stored, the fully metabolized forms are hydrophilic and readily excretable (Hodgson 1974, Malins 1977). Thus the process of excretion of HCs involves their conversion into more toxic intermediates. In fact it has been repeatedly stated that most substances generally referred to as carcinogens are in fact precarcinogens that must be bioactivatd by organisms before they can exert their carcinogenic effects (Miller 1970, Bend and others 1977, Oesch and others 1976, Jerina and Daley 1974).

When very high levels of chlorinated HCs are accumulated, the ability of organisms to partition them into lipid tissues and vacuoles may be exceeded (Allen and others 1976). At this point there is necrosis of tissue. It is apparent that in shellfish, petroleum and chlorinated HCs are accumulated in lipid tissue and that because of limited amounts of lipid there is a limited capacity to accumulate the pollutants; at a certain level of HC exposure, lipid tissue becomes HC saturated (Fong 1976).

It is well established that HCs are also accumulated in lysosomes (Allison 1969, Moore and others 1978a) that are rich in lipid and protein (Koenig 1969). When very high levels are accumulated within the lysosomes, they will lyse with release of hydrolytic enzymes and concurrent toxic effects due to autolysis (Moore and others 1978a, Amlacher 1970). In Mytilus edulis, petroleum HCs are known to be taken up by lysosomes in the digestive gland (Moore 1979), where lysosomal damage is induced.

There is a growing body of evidence that cytochrome P450-linked mixed-function oxidases responsible for the bioactivation of HCs exist in marine bivalves, including Mytilus edulis, despite previous evidence to the contrary (Khan and others 1972, 1974; Moore 1979). It is apparent that examination of tissues containing high levels of proteases, without the use of protease inhibitors, may produce incorrect results indicating that levels of cytochrome P450 are very low in marine organisms (Payne 1977). It is also apparent that the levels determined by various workers may be dependent upon the tissues examined (Khan and others 1972). Moore (1979) has found that the highest activities of cytochrome P450-linked reactions may exist in the blood cells of Mytilus edulis. In this context, Moore (1979) has suggested that the high incidences of neoplastic blood cells associated with aromatic HC contamination may be related to the ability of blood cells of Mytilus edulis to bioactivate these compounds. Furthermore, it is apparent that high levels of the enzymes responsible for deactivation of the bioactivated intermediates exist in Mytilus edulis (Bend and others 1977). Thus, the biochemical bioactivation and deactivation systems in Mytilus edulis appear to be similar to those found in other organisms (Harshbarger 1977, Bend and others 1977). Furthermore, the limited capacity of lysosomes and lipid-rich tissues to accumulate HCs suggests that overloading of these storage compartments may have toxic consequences similar to those in higher organisms.

Although Mytilus appears to have the metabolic capability eventually to detoxify petroleum HCs, it seems that the maximum capacity of the system is rather low. This is similar to the situation recorded in a crab by Burns (1976) where, although enhanced activity of detoxifying enzyme systems could be induced by exposure to HCs, even at maximum activity there was little impact on the total burden of accumulated HCs. This property is an advantage in a monitoring context, since physiological effects of the pollutant are still felt (and therefore measurable), whereas the biochemical response can act as a signal identifying the causative pollutant.

Both chlorinated and petroleum HCs have short half-lives in marine bivalves. Dunn and Stich (1976) have determined the half-life of petroleum HCs to be 16 days in Mytilus edulis, and Langston (1978) has found the half-life of PCBs in bivalves to be between 7 and 14 days. However, as found by Langston, the more highly chlorinated and hence more toxic chlorinated HCs may have a much longer half-life. Therefore, to dismiss bioactivation and deactivation processes in Mytilus edulis as unimportant because of the high rate of excretion of these compounds ignores the possibility that the more toxic components may not be excreted as quickly as some less toxic components. More research is needed in Mytilus to determine which components of HCs are retained longest and their potential to exert both toxic and carcinogenic effects. More complex HCs, such as benzopyrene and dimethylbenzanthracene, are retained in mammalian lysosomes for much longer periods than are simpler HCs such as anthracene (Allison and Malluci 1964).

Moore and others (1978a) have indicated that measures of the lysosomal latency of Mytilus edulis serve to quantifiably assess the degree of petroleum HC pollution. This measure of latency is based indirectly upon an assessment of the loading of lysosomes with petroleum HCs in order to partition them away from sensitive cellular macromolecules. However, since lysosomes accumulate other pollutants (Moore 1977) and are known to destabilize and subsequently disrupt with exposure to other pollutants, this measure does not provide an assessment of the specific effects of HCs. For this reason, it is suggested that future research involve an assessment of the partitioning of HCs between their storage detoxification systems (lysosomes, and lipid vacuoles and other lipid-rich tissue components) and their sites of toxic action (for instance, sensitive components of the cytoplasm and nucleoplasm such as enzymes, DNA, and RNA).

Further, an evaluation is needed of the role of bioactivation in the toxicology of these pollutants in Mytilus edulis. In this regard, it can be argued that the ratio of the activity of the bioactivation enzymes (cytochrome P450 system) to the deactivation enzymes (epoxide hydrases and glutathione-s-epoxide transferases) may be the biochemical measure indicative of whether or not a HC will exert toxic or carcinogenic effects (Jerina and

Daley 1974, Oesch and others 1976, Bend and others 1977).
It has been demonstrated in some organisms that the level of
carcinogenicity of some compounds depends directly upon the
levels of glutathione available to conjugate with the
bioactivated intermediates (Maugh 1978b).

## Radionuclides

Radionuclides have been implicated as responsible for
damage to cell organelle membranes and macromolecules,
particularly chromosomes (Templeton and others 1978). As
with other pollutants, disruption of lysosomes has been
implicated as a major vehicle for toxic effects of
radionuclides (Allison 1969). In contrast with trace metals
and HCs, there appears to be no effective detoxification
system for radionucides. Therefore, it does not appear
possible to measure the partitioning of radionuclides
between a detoxification system and its site of toxic action
as it is with other pollutants. In this context it has been
stated that there is no threshold level of radionuclides at
which toxic effects become apparent (Roberts 1976).
Adaptation to radionuclides appears to consist of increases
of the enzymes responsible for cellular repair processes
(Mix 1976). Therefore, we suggest that any attempts to
quantify the biological effects of radionuclides might
involve assessments of increases of cellular repair enzymes,
lysosomal latency, and chromosomal damage.

## Interactions Between Pollutants

Perhaps the most unexplored and yet important area of
toxicology is that of interaction between pollutants in
regard to the effects of detoxification. The following
examples provide indications of the kinds of interactions
that can occur between pollutants. Although they are not
specific for mussels, the principles demonstrated here may
have applications to mussels since it appears that the basic
detoxification systems of mussels are similar to those in
other organisms.

It is apparent that deficiencies of an essential metal
may increase the toxicity of a toxic metal. It has been
found in ducks that have an apparently zinc-replete enzyme-
containing pool that almost all cadmium is bound to
metallothionein and thereby detoxified. In those with low
levels of zinc in the enzyme-containing pool, almost all

cadmium appeared in the enzyme-containing pool, presumably because it was more effectively able to compete for zinc binding sites in metalloenzymes (Brown and Chatel 1978a). Thus, in zinc-deficient animals, cadmium is more effectively able to exert toxic effects, since it is not as readily detoxified by metallothionein. The protective effect of zinc on cadmium toxicity is well established (Flick and others 1971) and a consideration of both cadmium and zinc levels in regard to detoxification of cadmium provides valuable clues as to the actual mechanism of this protective effect.

Organic pollutants may be able to affect the zinc status of organisms and thus their ability to detoxify cadmium. Organic carcinogens have been shown to cause zinc deficiency in organisms (Fong and others 1978, Brown and Chan 1978, Brown 1978). In particular, these losses of zinc have been shown to occur from the enzyme-containing pool (Brown and Chan 1978, Brown 1978). When cadmium is given alone, almost all cadmium accumulates on metallothionein. When cadmium is given with an organic carcinogen, almost all cadmium accumulates in the enzyme-containing pool, presumably due to the carcinogen-induced zinc deficiency in the enzyme-containing pool (Brown and Chatel 1978b). In this context, it is interesting to note that both cancerous fish and humans have increases of cadmium and the cadmium/zinc ratio in the enzyme-containing pool (Brown 1977, Brown and Knight 1978). In cancerous organisms, almost all cadmium occurs in the enzyme-containing pool, whereas in noncancerous organisms, almost all is detoxified by metallotheionein. Of relevance may be the fact that the enzymes controlling cell division are zinc-containing enzymes (e.g., DNA polymerase, RNA polymerase, thymidine kinase) (Vallee 1976). Furthermore, zinc has been repeatedly shown to prevent cancer from developing when given with carcinogens (Gunn and others 1963, 1964; Poswillo and Cohen 1971; Ciapparelli and others 1972; Fong and others 1978). For these reasons, it is suggested that particular attention be paid to the cytoplasmic levels and distributions of cadmium and zinc in areas thought to be polluted with carcinogens and where carcinoma in mussels is known to occur. Perhaps an evaluation of zinc levels in various carcinogen-polluted areas will provide clues as to why cancers occur in some of these areas and not in others.

Of interest also is the influence of cadmium and mercury on cytochrome P450. In several studies, both of these metals have been shown to result in cessation of activity of this enzyme system (Yoshida and others 1976, Johnston and others 1974, Carlson 1975, Pani and others 1976). This may have important implications for both the bioactivation and detoxification of petroleum and chlorinated HCs.

In summary, use of the above methods may provide means to determine which pollutants are present at higher than threshold levels for toxic effects to the organism. When several pollutants are present, only measures of their partitioning between detoxification systems and sites of toxic action will provide clues as to which pollutants are biologically available inside organisms to cause harm to the organisms. No matter what the levels of pollutants in an organism, only those present at the site of toxic action will be able to cause direct toxic effects. Those that are detoxified will be a health hazard only in that they are stressful; this is because they require a mobilization of energy reserves so that the detoxification systems can be synthesized. To assess these effects we recommend careful monitoring both of indices that are indicative of the general stress syndrome and of the partitioning of pollutants between their detoxification systems and their sites of toxic action.

## CONCLUSIONS AND RECOMMENDATIONS

As noted earlier, any consideration of the biological effects of pollution becomes a consideration of stress and its symptoms in the organism. Our first recommendation is that the application of biological techniques to monitoring of animal health be done strictly in the context of our definition of stress; the alternative is an involvement in biological measurements that may prove to be trivial in terms of understanding pollution effects.

Within this definition of stress we identify the general stress syndrome of bivalves in response to environmental change. Elements of this stress syndrome are immediately available for involvement in mussel watch monitoring programs and we recognize the following as showing particular potential:

1. the scope for growth, representing a fundamental adaptive response;
2. condition indices that point to alterations in the nutritive status of the animal;
3. the oxygen/nitrogen ratio, since it provides an integrated physiological measurement of alteration in the balance between catabolic processes;

4. the energy charge ratio, since this registers a fundamental metabolic consequence of stress;

5. the latency of lysosomal enzymes, which is a basic element of the general cytotoxic response in mussels;

6. ratios between certain amino acids, particularly the taurine/glycine and serine/threonine ratios;

7. the incidence of granulocytomas as indicative of cellular response to injury, and the incidence of hemocytic infiltration of tissues;

8. the incidence of neoplasm, particularly the gonadal and the hemopoeitic neoplasms;

9. pathological damage to gill, kidney, and digestive gland tissue, particularly the alteration in digestive rhythm in the digestive gland; and

10. the incidence of parasites and the degree of inflammatory host response to parasitic infestation.

If all these indices of stress cannot be adopted at once, we recommend that condition index (2), lysosomal latency (5), and the cytological indices (7, 8, and 9) be considered, since indices 5 and 7 to 9 all involve similar cytological screening techniques, and index 2 is well within the purview of most laboratories to measure. Where possible, scope for growth (1) should also be measured.

Having recognized that general stress indices are now available, we wish to point to the need for indices that are more specifically responsive to particular classes of contaminants. Some of these have been discussed in this chapter, but more research is required before we can recommend their inclusion in monitoring programs. We urge that this research be conducted within the framework of ideas explored earlier; namely the identification of sites of detoxification and sites of toxic action. Only when we can discriminate between these two, and can ascribe maximum capacities to the former, can we begin to establish very precise relationships between contaminant concentrations and biological effect.

In addition, we recommend that both chemists and biologists be encouraged to perform research on the combined actions of pollutants, particularly pollutants of different kinds. There is already some evidence of true synergistic (or supra-additive) effects between pollutants, but much more research is needed in this very important area.

In seeking to provide some further guidelines for research on this topic of specific stress indices we identify three areas of promise:

1.  the various cellular responses (e.g., differential hemocytic response) to metal contamination;

2.  the role of metallothioneins in detoxifying metals; and

3.  the role of the cytochrome P450-linked mixed-function oxidases in detoxifying organic pollutants.

We consider that some of these biological procedures should be incorporated into mussel watch programs now. However, monitoring biological effects is labor intensive and likely to be expensive. We therefore recognize that, in the first instance, biological involvement is likely to be confined to pollution hot spots. In these areas the biologist will hope to establish dose-response relationships for biological effects. These relationships are likely to be complex and they are unlikely to resemble the dose-response relationships of the laboratory bioassay. Nevertheless, perhaps with the use of simulation modeling techniques, these complex relationships need to be quantified. The biologist may then become involved in relatively less contaminated areas in attempts to build up a more complete picture of the effects of known environmental levels of contamination.


APPENDIX 6-A

CALCULATION OF THE SCOPE FOR GROWTH

In this appendix we illustrate how the scope for growth is calculated from data obtained in the field for rates of oxygen consumption, rates of filtration, absorption efficiency, and the abundance and calorific value of particulate material available as food. The data are taken from a study by scientists at Plymouth, United Kingdom, of Mytilus edulis at Kings Dock in Swansea, South Wales. Our aim is not to detail how the various measurements were made, but rather to show the form of the statistical analysis of the data and its integration as the scope for growth. We refer to a forthcoming manual on techniques for measuring the effects of stress in bivalves (Bayne and others, in preparation) for a description of techniques.

1.  Rate of Oxygen Consumption. The data are filed in the following format:

*Date*: 26, 27, 28 May

| Dry weight (g) | O$_2$ Consumption Rate (ml h$^{-1}$) | Salinity | Time of Day |
|---|---|---|---|
| 1.135 | 0.599 | 30.5 | 1300 |
| 1.133 | 0.919 | 30.5 | 1345 |
| 0.986 | 0.720 | 30.5 | 1050 |
| 1.428 | 1.015 | 30.5 | 1150 |
| 0.556 | 0.770 | 30.5 | 1235 |
| 0.542 | 0.471 | 30.5 | 1340 |
| 0.579 | 0.681 | 30.5 | 1425 |
| 3.059 | 1.180 | 30.5 | 1530 |
| 1.017 | 0.805 | 30.0 | 1120 |
| 0.898 | 0.625 | 30.0 | 1215 |
| 0.282 | 0.266 | 30.0 | 1320 |
| 1.098 | 0.855 | 30.0 | 1420 |
| 1.683 | 1.383 | 30.0 | 1510 |
| 1.130 | 1.239 | 30.0 | 1550 |

These data are then transformed by log$_{10}$ and analyzed by least squares linear regression with the following results (where Y is transformed oxygen consumption and X is transformed dry-flesh weight):

Mean $Y$ = -0.1170    Mean $X$ = -0.2026 $E$-1

| Source | Dry Flesh | Sums of Square | Mean Square | F |
|---|---|---|---|---|
| Regression | 1 | 0.313 | 0.313 | 28.0 |
| Residual | 12 | 0.134 | 0.112 E-1 | |
| Total | 13 | 0.447 | | |

Correlation coefficient = 0.84

Standard deviation of $Y$ on $X$ = 0.106

This is repeated for all the data from any one population; then the regressions from each date are compared by covariance analysis to establish the significance of differences between the parameter values ("slope" and "intercept") in the different regression equations. For example, following six visits to the Kings Dock site (over a period of 12 months), the final analysis of covariance was:

| Source | Dry Flesh | Sums of Square | Mean Square | F |
|---|---|---|---|---|
| Total Residual | 56 | 0.8396 | 0.149 E-1 | |
| Difference of slopes | 5 | 0.297 E-1 | 0.595 E-2 | 0.40 |
| Subtotal | 61 | 0.8694 | 0.142 E-1 | |
| Difference of Intercepts | 5 | 2.4295 | 0.486 | 34.09 |
| Total | 66 | 3.2988 | | |

This indicates no significant differences between slopes (the exponent values in the regression equation), but a highly significant difference between intercepts (equivalent to the rate of oxygen consumption by animals of 1 gram dry-flesh weight). The latter finding is due to seasonal variability in the intensity of metabolism.

2. Rate of Filtration. These data are filed and analyzed in the same way as the oxygen uptake measurements. For the data set under consideration the regression equation for the $\log_{10}$ transformed data is:

$$Y = 0.1034 + 0.675X$$
$$+/-\ 0.148\ E-1\ +/-\ 0.157$$

3. Absorption Efficiency. This is calculated according to Conover (1966).

| | Dry weight (mg):a | Ash-free dry weight (mg):b | b:a |
|---|---|---|---|
| Food | 2.02 | 1.07 | 0.53 = F |
| Feces | 49.42 | 13.17 | 0.27 = E |

Efficiency = $\frac{F - E}{(I-E)F}$ .100 = 67%

4. <u>Caloric Content of the Food</u>. This is determined from the ash-free, dry-weight measurements of suspended particulate material collected as a number of discrete samples during the period of the site visit to the population. The constant used in converting the ash-free dry weight to calories varies between sites and must be determined separately (see Bayne and Widdows 1978). For the data set under consideration:

| Ash-free dry weight of food (mg $l^{-1}$) | Calorific conversion constant (cals. $mg^{-1}$) | Calories available as food (cals. $l^{-1}$) |
|---|---|---|
| 1.26 | 3.8 | 4.79 |

5. <u>The Scope for Growth</u>. The rate of oxygen consumption (normalized for an animal of 1 gram dry-flesh weight) is converted to a caloric equivalent by multiplying ml $O_2$ $h^{-1}$ x 4.8. This value is then subtracted from the caloric equivalent of the absorbed ration, itself the product of rate of filtration (1 $h^{-1}$) x food available (cals $l^{-1}$), corrected for absorption efficiency. The result of these calculations is shown in Table 6-A.1.

APPENDIX 6-B

PREPARATION OF AQUATIC ANIMALS FOR
HISTOPATHOLOGICAL EXAMINATION

The information in this appendix is intended to apply to specific problems encountered in collecting and processing aquatic animals for routine histopathological examination. The text is adapted from a fuller account to be published by the U.S. Environmental Protection Agency (EPA). General histological techniques, described in the books listed at the end of this appendix, have been adapted by the Histology Unit of the Environmental Research Laboratory, Narragansett (ERLN), Rhode Island, for use with aquatic animals.

TABLE 6-A.1  Results of Calculations of Scope for Growth[a]

| Rate of Filtration ($l\ h^{-1}$) | Food Available ($cals.\ l^{-1}$) | Ingested Ration ($cals.\ h^{-1}$) | Absorption Efficiency | Absorbed Ration ($cals.\ h^{-1}$) | Respiratory Loss ($cals.\ h^{-1}$) | Scope for Growth ($cals.\ h^{-1}$) |
|---|---|---|---|---|---|---|
| $F$:1.27 | $f$:4.79 | $F \times f = C$ 6.081 | $e$ 0.67 | $C \times e = A$ 4.074 | $R$:3.773 | $A - R$ 0.301 |

[a]The result is a positive scope for growth of 0.301 cals. 1 h$^{-1}$ signifying some surplus energy available for growth and the production of gametes.

## Collection of Animals

When collecting aquatic animals, every effort should be made to keep physical damage to a minimum. Animals that grow attached to rocks or other objects may be carefully cut or scraped from their point of attachment with a sharp knife or spatula and put directly into a collecting container. If the animals are not easily removed, it may be possible to bring the whole object with the animals attached back to the laboratory and totally immerse it in fixative. After fixation the animals are more easily removed from their object of attachment.

Hand collecting or trapping of bottom organisms is preferred to dredging or dragging because the latter methods may cause sand or other debris to become embedded in the tissue of the animals; also, shells may be broken and the animals crushed. These animals are not desirable for histopathological examination because of the damage done to the tissues and of the difficulties debris causes when sectioning the animals. If the animals have extensive abrasions or hemorrhages due to the collecting process, it is suggested that they not be used. When collecting from an area that has mass mortalities, such as in a fish kill or oil spill site, one should try to collect animals that show some signs of life. Autolysis, which brings about post mortem changes, begins in some tissues while the animal is still moribund and thus makes accurate analysis difficult.

## Observations

Observations should be made when preparing the animals for fixation. If the animals are fixed in the field, notes should be taken and the information logged when the animals get to the laboratory. Date, species, area of collection, type of experiment, and other data should be recorded either on a standardized form or on sheets developed particularly for the species involved. A log book should be kept for this information and to give each animal its own identification number.

The following are suggestions of things to look for while trimming the animals: (a) malformation, especially in the gills and body mass; (b) parasites, both external and internal; (c) firmness of tissue; (d) changes in the color of the internal organs such as the digestive diverticular and gonadal tissue; and (e) cyst and tumor formation.

## Fixatives

The most important step in the preparation of aquatic animals for histopathologic examination is the process of fixation.  Fixation is necessary to prevent autolysis and to preserve the tissue in as lifelike a condition as possible. Adequate fixation with the proper fixative is critical for the preservation of the cellular inclusions and secretions as well as the normal cell structure of aquatic animals. Helly's fixative (Jones 1964) made with zinc chloride instead of mercuric chloride is the fixative of choice for most invertebrates (Barszcz and Yevich 1976).  Helly's is an excellent cytological fixative for aquatic invertebrates, preserving the nucleoplasm of the nuclei of ova and the granules of the secretory cells and amoebocytes better than other fixatives tested.  Tissues fixed in Helly's stain intensely with hematoxylin and eosin and many special stains.

Dietrich's fixative (Gray 1954) gives good cellular detail and there is little shrinkage or distortion of the cells.  Animals can be stored in Dietrich's for 2 to 3 months, if necessary, without excessive hardening of the tissues.  The major drawback to using Dietrich's or any other fixative that contains glacial acetic acid for invertebrates is that it appears to cause the loss of secretions and granules and loss of nucleoplasm from ova.

Ten percent buffered Formalin is seldom used because it causes shrinkage, loss of cellular detail, hardening of the tissues, and poor staining qualities.  It and other fixatives noted in the literature are used only as required for special stains.

Formulas for Helly's and Dietrich's fixatives are given below.


## Helly's Fixative

1,000 gm zinc chloride*
500 gm potassium dichromate
20 l distilled water
Add 5 ml of 37 to 40 percent formaldehyde per
100 ml of fixative at time of use.

*    Zinc chloride is used in Helly's fixative instead of mercuric chloride because it is less toxic and does not have a precipitate that must be removed during staining.  It has the same fixing and mordant qualities as mercuric chloride.

## Dietrich's Fixative

9,000 ml distilled water
4,500 ml 95 percent ethanol
1,500 ml 40 percent formaldehyde
300 ml glacial acetic acid


Tissues must be immersed in at least 20 times their
volume of fixative.  In the laboratory, standard glass jars
are used.  When animals cannot be brought back to the lab
for fixation within a reasonable amount of time, they must
be fixed in the field.  Plastic bottles of the proper sizes
are recommended for field work, as described in the
following sections.  Animals to be fixed in Dietrich's are
trimmed and stored in these plastic containers in fixative
until processed.

### Preparation of Tissues for Processing

Acceptable slides cannot be produced unless the animals
are properly prepared for processing.  Animals are killed by
direct immersion in fixative.  Most bivalves must be removed
from their shells before immersion.  After 10 to 15 minutes
in fixative to allow firming of the tissues, the animals are
removed and trimmed as detailed in the following section.
Immediately after trimming, the tissue is returned to
fixative.

Hard-shell clams, such as the black clam (_Arctica
islandica_) and the quahog (_Mercenaria mercenaria_), are
opened by cutting the ligament and wrenching the shell apart
to tear the muscle.  With a quahog knife, the animal is
shucked free of both valves and dropped into fixative.  Care
should be taken to prevent the tearing of the mantle when
removing the animals.  Animals such as the soft-shell clam
(_Mya arenaria_) and the surf clam (_Spisula solidissima_) can
be opened by inserting the quahog knife into any available
opening between the mantle and the valve.  Oysters are
opened by inserting an oyster knife into the ligament and
wrenching the shells apart with a twist of the wrist.  A
quahog knife or a spatula is used to loosen the mantle and
muscle from the valves.  Scallops are opened by inserting a
spatula between the mantle and valve and cutting the muscle
from the valve with a sweeping motion.  Mussels are opened
by forcing a sharp quahog knife between the two valves
slightly above the ligament so that the tip of the knife is

between the mantle and shell. With a sweeping motion, the muscle and mantle are cut loose from the shell.

After the animals have been removed from their shells, they are put in fixative for 10 to 15 minutes, then removed and trimmed. Almost all large animals of the clam variety are cross-sectioned, beginning at the mouth, into 3 to 4 mm sections. Oysters are cut in half through the body mass just to the side of the pericardial cavity opposite the adductor muscle; then each half is cut sagittally. All scallops are cut sagittally into two, three, or four sections depending on the size of the animal. Mussels are trimmed by separating the mantle from the body mass and cutting the body mass sagittally.

Small bivalves are dropped into fixative after their shells have been cracked slightly, but not crushed. This allows the fixative to penetrate and the animals will be easier to remove after fixation. They are cut sagittally or, if very small, processed whole. If the spat are too small to open and remove from their shells after the appropriate fixation time, the fixative is removed and Decal added for decalcification. Decalcification is complete when no hard areas of shell remain.

## Final Trimming

The final trimming on all large pieces of tissue is done after fixation and decalcification are completed. The tissue is cut into sections no more than 3 mm thick and small enough in diameter to fit into the cassettes and eventually into the embedding molds. The surface of tissue to be examined is placed face down in the cassette to enable the embedder to orient the tissue in the paraffin block correctly. The animal's identification number is included in each cassette on a small piece of paper. All tissue is washed overnight (16 hours) in a running water bath to remove all excess fixative. If the tissue has been in Decal, it must be washed for at least 24 hours to remove the acid. After washing, the cassettes are stored in a dehydrating solution until processed.

## Processing the Tissue

Tissues are prepared for embedding in paraffin by a series of dehydration and clearing steps. Graded alcohols can be used for dehydrating, and xylene or chloroform for clearing the tissue and preparing it for infiltration by the paraffin. Dioxene alone can be used to both dehydrate and clear the tissue. Technicon's dehydrating and clearing reagents, S-29 and UC-670, respectively, are used at ERLN.

Many of the histological technique books give examples of dehydrating, clearing, and infiltrating schedules. The schedules we have found most useful with Technicon products are listed in Table 6-B.1. These schedules can be used for hand or machine processing of tissue. The 16-hour or night run is convenient to use for machine processing of most tissues. For hand processing on the 16-hour schedule, the cassettes can be left in fresh S-29 overnight and the schedule completed the next day.

Tissues that have been stored in a dehydrating solution for several days can be processed on the day run. This schedule can also be used for soft tissues that are easily penetrated by the reagents. Penetration by the paraffin is enhanced by vacuum infiltration for at least half an hour. This also helps to remove air bubbles trapped in the tissue. The short run is used for larvae, eggs, nerve fibers, and ganglia, and any other delicate tissues. These tissues should not be put into a vacuum infiltrator.

## Sectioning

The amount of sand and shell particles found in many molluscs and other invertebrates makes the use of regular microtome blades impractical. Disposable knives are used on a rotary microtome to cut sections at 6 μm for routine work. The tissue ribbons are floated on a warm water bath to allow the paraffin to stretch and the wrinkles to be removed. One, two, or more sections, depending on the size of the block, are mounted on 1 x 3 inch etched slides that have been marked with the animal's identification number. The slides are drained, put into a staining rack, and held in a 57°C oven for at least 3 to 4 hours until the protein in the tissue adheres to the slide. With this treatment, no adhesive such as gelatin or albumen is usually necessary. After being in the oven, they can be stained or stored in slide boxes at room temperature until needed. Usually one slide is cut from each block. Then the block is sealed in paraffin to prevent dehydration and filed for future reference. If upon examination, something of particular interest is found on the slide, the block can be recut and special stains done as necessary.

TABLE 6-B.1  Routine Processing Schedules

| Beaker No. | Night Run Reagent | Time (hours) | Beaker No. | Day Run Reagent | Time (hours) | Beaker No. | Short Run[a] Reagent | Time (minutes) |
|---|---|---|---|---|---|---|---|---|
| 1 | S-29 | 1 | 1 | S-29 | .5 | 1 | S-29 | 10 |
| 2 | S-29 | 2 | 2 | S-29 | .5 | 2 | S-29 | 10 |
| 3 | S-29 | 2 | 3 | S-29 | .5 | 3 | S-29 | 10 |
| 4 | S-29 | 2 | 4 | S-29 | .5 | 4 | S-29 | 10 |
| 5 | S-29 | 2 | 5 | S-29 | .5 | 5 | S-29 | 10 |
| 6 | S-29 | 2 | 6 | S-29 | .5 | 6 | S-29 | 10 |
| 7 | UC-670 | .5 | 7 | UC-670 | .5 | 7 | UC-670 | 10 |
| 8 | UC-670 | 1.5 | 8 | UC-670 | 1 | 8 | UC-670 | 10 |
| 9 | Paraffin | 1 | 9 | Paraffin | 1 | 9 | Paraffin | 10 |
| 10 | Paraffin | 1 | 10 | Paraffin | 1 | 10 | Paraffin | 10 |
| 11 | Paraffin | .5 | 11 | Paraffin | .5 | 11 | Paraffin | 10 |
| Vacuum | Paraffin | .5 | Vacuum | Paraffin | .5 | Embed as quickly as possible | | |

[a] Delicate tissues

## Staining

For routine microscopic examination, slides are stained
with Harris' hemotoxylin and eosin. The time spent in each
of the stains is determined by the type of fixative used and
whether or not the tissue has been in Decal. Helly's-fixed
tissue stains more quickly and intensely than Dietrich's-
fixed tissue. Decalcifying the tissue increases the time
required for the staining reactions to take place.

When particular cell components are to be examined,
special stains are used. Heidenhain's Azocarmine-Aniline
Blue is used to stain the granules in the neurons of
molluscs. To facilitate the staining of the granules, the
slides are held in the Azocarmine solution in the oven for 1
hour. The Heidenhain's method is also used to stain the
granules found in the digestive diverticula of molluscs. In
this case, the normal staining time is sufficient.

Movat's Pentachrome stain is used to permit easy
identification of various cell types, particularly mucous
secretory cells. This stain is modified by leaving out the
elastic portion of the procedure to save time. McManus' PAS
or Best's carmine stains are used for glycogen
determination. Fat deposition is best demonstrated by the
osmium tetroxide method for fat in paraffin sections
developed by the Armed Forces Institute of Pathology.
Dahl's Alizarin Red S method is used for staining calcium
deposits.

After the slides are stained, they are coverslipped
using Harleco's Synthetic Resin and No. 1-1/2 24 x 60 mm
coverslips and labelled with a permanent tag bearing the
laboratory name and the animal's identification number. The
number of slides from each animal and the date the slides
are completed are recorded in the log book. After the
slides have been examined by the pathologist, they are filed
for future reference.

## APPENDIX 6-C

## AN INTER-LABORATORY HISTOPATHOLOGICAL
## COMPARISON OF U.S. MUSSEL WATCH MATERIAL FROM TEN SITES

Material from 10 sites, sampled in 1976, was assessed at
the Narragansett laboratory (Histopathology Unit, U.S.
Environmental Protection Agency) by Yevich and Barszcz
(1977) and sent under a different coding to Plymouth, U.K.

(National Environmental Research Council Institute for Marine Environmental Research) for independent assessment by Lowe and Moore (1978). Criteria were as described in Chapter 6, and the results are listed below, the first assessment summarizing the findings of Yevich and Barszcz (Y.B.) and the second those of Lowe and Moore (L.M.) (Tables 6-C.1 and 6-C.2).

TABLE 6-C.1  Assessment

---

A.  Manhasset Neck Station (MN 761009) M. edulis

Y.B.  Granular kidneys.  Poor digestive diverticular. Some amoebocytic infiltration in connective tissues, byssus gland, and pericardial cavity.  Increased secretory activity in eight animals; hemopoietic tumor in one animal.  VERY POOR

L.M.  There are six cases of granulocytomas, one elevated hemocyte count, ten cases of digestive cell transformation, and one response to parasitism.  POOR

B.  Block Island Station (BLK 761001) M. edulis

Y.B.  Extensive Bucephallus infestation and microsporidia in gonad.  Parasitic destruction of digestive tubules, with amoebocytic reaction in one animal. Pericardial gland necrosis in one animal.  VERY POOR

L.M.  Although the digestive rhythm appears to be normal, this group exhibits seven cases of granulocytomas, four animals have extensive infiltrations, and three show host responses to parasites.  One animal exhibits cell masses in the pericardial gland.  VERY POOR

C.  Willapa Bay Station (WB 760804) M. edulis

Y.B.  Considerable protozoan parasite invasion of the digestive cells, some amoebocytic infiltration. Considerable variation in reproductive condition with some gamete resorption.  POOR

**L.M.** The absence of any irregularity must rate these
VERY GOOD


D. North San Francisco Bay Station (NSF 760701) **M. edulis**

**Y.B.** Poor and varied development of the reproductive
tract. Three with hemopoietic neoplasms, with invasion of
gonadal and digestive tissues. POOR

**L.M.** The digestive rhythm appears normal; there are,
however, three cases of hemocytic infiltrations, no gill
abnormalities, and no granulocytomas. GOOD


E. Point La Jolla Station (LJ 760920) **M. californianus**

**Y.B.** Healthy kidney structure, normal digestive cell
structure. Some swelling of gill inter-lamella. GOOD

**L.M.** Although the digestive rhythm does vary, the
general picture is good; there are no tissue parasites and
no granulocytomas or hemocytic infiltrations. GOOD


F. Millstone, Connecticut Station (MIL 761004) **M. edulis**

**Y.B.** Poor reproductive development with some
microsoporidial infestation. Granulocytomas in byssus
gland. Five animals with calcium concentrations in kidney.
Parasitic incidence medium. POOR

**L.M.** The digestive rhythm is severely affected with
extensive tubule epithelial cell transformation; there are
two cases of hemocytic infiltrations and two granulocytomas.
Three animals have parasites, of which one initiated a host
response. Gills are all normal. POOR


G. Gold Beach Station (GB 760720) **M. californianus**

**Y.B.** Animals in healthy condition, no abnormalities.
VERY GOOD

**L.M.** The absence of any irregularity must rate this
group VERY GOOD.

H.   Cape Ann Station (CA 760921) M. edulis

Y.B.  Some granulocytomas (byssus gland) and two animals
with amoebocytic infiltration in muscles.  Some increase in
mucous secretory activity in gills.  Tissue alterations not
extensive, however.  GOOD

L.M.  The digestive rhythm varies; however, of the ten
mussels only three exhibit epithelial changes to any marked
degree.  Two animals have hemocytic infiltrations; no gill
abnormalities and no granulocytomas are apparent.  FAIR


I.   San Diego Station (SDH 760425) Mixed M. edulis, M.
     californianus

Y.B.  Some amoebocytic infiltration in reproductive
tissue, some microsporidial infestation.  Normal kidney and
digestive gland structure, but with some amoebocytic
infiltration.  Five animals with parasitic inflammatory
response.  GOOD

L.M.  Tubule epithelial cells are in good conditions,
there are three cases of increased hemocyte activity, the
oviducts are occluded in one animal with disruption of
follicles, there are no granulocytomas.  FAIR


J.   Humboldt (H 760714) M. californianus

Y.B.  Three animals with kidney calcium concretions.
All animals had light to heavy swelling of the gill inter-
lamellar connective tissue.  Zero parasitic infestation.
FAIR

L.M.  In the absence of any tissue irregularity this
group must be rated VERY GOOD (see Table 6-C.2).

---

The agreement between these two independent assessments
is fair.  The greatest differences refer to stations C, D,
and J.  For the first two of these, Yevich and Barszcz rely
heavily on the condition of the reproductive tract and
occurrence of gamete resorption.  Lowe and Moore comment
that they do not use these criteria because of the
variability associated with the spawning cycle; naturally

TABLE 6-C.2   Ranking Comparison

| Assessment | Y.B. | L.M. |
|------------|------|------|
| Very Good | G | C, G, I |
| Good | E, H, I | D, E |
| Fair | J | H, I |
| Poor | C, D, F | A, F |
| Very Poor | A, B | B |

spawned animals often have the appearance of great tissue disruption but, they argue, this is normal. For population J, Yevich and Barszcz point to the high incidence of inter-lamellar connective tissue swelling in the gills. Lowe and Moore, who are relatively unfamiliar with M. californianus, did not regard this as a pathological condition.

Further comparisons of this kind should be encouraged. For our present report we have not suggested using the condition of the reproductive tract as a pathological criterion, other than conditions of heavy parasitic infestation of the gonad, particularly with microsporidia, for the reasons given above.

## REFERENCES

Allen, J.R., L.A. Carstens, and D.A. Barsotti (1974) Residual effects of short-term, low-level exposure of nonhuman primates to polychlorinated biphenyls. Toxicol. Appl. Pharmacol. 30:440-451.

Allen, J.R., L.A. Carstens, L.J. Abrahanson, and R.J. Marlar (1975) Responses of rats and nonhuman primates to 2,5,2',5'-tetrachlorobiphenyl. Environ. Res. 9:265-273.

Allen, J.R., L.A. Carstens, and L.J. Abrahanson (1976) Responses of rats exposed to polychlorinated biphenyls for fifty-two weeks. I. Comparison of tissue levels of PCB and biological changes. Arch. Environ. Contam. Toxicol. 4:404-419.

Allison, A.C. (1969) Lysosomes and cancer. Pages 178-204, Lysosomes in Biology and Pathology, Vol. 2, edited by J.T. Dingle and H.B. Fell. Amsterdam: North Holland-American Elsevier.

Allison, A.C. and L. Malluci (1964) Uptake of hydrocarbon carcinogens by lysosomes. Nature 209:303-304.

Allison, A.C. and G.R. Paton (1965) Chromosome damage in
human diploid cells following activation of lysosomal
enzymes. Nature 207:1170-1173.
Amlacher, E. (1970) Textbook of Fish Diseases, translated by
D.A. Conroy and R.L. Herman. T.H.F. Publications.
Anderson, E.P. (1975) Condition and Zinc Content of Oysters
in the Crofton Area, 1975. A report submitted to British
Columbia Forest Products, June 20, 1975.
Anderson, J.W., J.M. Neff, and S.R. Petrocelli (1974)
Sublethal effects of oil, heavy metals and PCBs on
marine organisms. Pages 83-121, Survival in Toxic
Environments, edited by M.Q. Khan and J.P. Bederka, Jr.
New York: Academic Press.
Atkinson, D.E. (1977) Cellular Energy Metabolism and Its
Regulation. New York: Academic Press.
Barry, M. and P.P. Yevich (1975) Ecological, chemical and
histopathological evaluation of an oil spill site: III.
Histopathological study. Mar. Pollut. Bull. 6:171-173.
Barszcz, C.A. and P.P. Yevich (1976) Preparation of copepods
for histopathological examination. Trans. Am. Microsc.
Soc. 95(1):104-108.
Barszcz, C.A., P.P. Yevich, L.R. Brown, J.D. Yarbrough, and
C.D. Minshew (1978) Chronic effects of three crude oils
on oysters suspended in estuarine ponds. J. Environ.
Pathol. Toxicol. 1:879-895.
Bayne, B.L. (1975) Aspects of physiological condition in
Mytilus edulis L. with special reference to the effects
of oxygen tension and salinity. Pages 213-238,
Proceedings 9th Europ. Mar. Biol. Symp., edited by H.
Barnes. Aberdeen: Aberdeen University Press.
Bayne, B.L., ed. (1976) Marine Mussels: Their Ecology and
Physiology. Cambridge: Cambridge University Press.
Bayne, B.L. and C. Scullard (1977) Rates of nitrogen
excretion by species of Mytilus (Bivalvia: Mollusca). J.
Mar. Biol. Assoc. (U.K.) 57:355-369.
Bayne, B.L. and R.J. Thompson (1970) Some physiological
consequences of keeping Mytilus edulis in the
laboratory. Helgo. Wiss. Meeresunters. 20:528-552.
Bayne, B.L. and J. Widdows (1978) The physiological ecology
of two populations of Mytilus edulis L. Oecologia
37:137-162.
Bayne, B.L., P.A. Gabbott, and J. Widdows (1975) Some
effects of stress in the adult on the eggs and larvae of
Mytilus edulis L. J. Mar. Biol. Assoc. (U.K.) 55:675-
689.

Bayne, B.L., D.R. Livingstone, M.N. Moore, and J. Widdows (1976) A cytochemical and a biochemical index of stress in Mytilus edulis L. Mar. Pollut. Bull. 7:221-224.

Bayne, B.L., D.L. Holland, M.N. Moore, D.M. Lowe, and J. Widdows (1978) Further studies on the effects of stress in the adult on the eggs of Mytilus edulis. J. Mar. Biol. Assoc. (U.K.) 58:825-841.

Bayne, B.L., M.N. Moore, T.H. Carefoot, and R.J. Thompson (1979a) Cells and internal function of the hemolymph and its responses to tissues implants in Mytilus californianus (Conrad) (in preparation).

Bayne, B.L., M.N. Moore, J. Widdows, D.R. Livingstone, and P. Salkeld (1979b) Measurement of the responses of individuals to environmental stress and pollution. Philos. Trans. Roy. Soc., Ser. B (in press).

Bend, J.R., M.O. James, and P.M. Dansette (1977) In vitro metabolism of xenobiotics in some marine animals. Ann. N.Y. Acad. Sci. 298:505-521.

Bernhard, F.R. (1973) Crystalline style formation and function in the oyster Crassostrea gigas (Thunberg, 1795). Ophelia 12:159-170.

Blaylock, B.G. and J.R. Trabalka (1978) Evaluating the effects of ionizing radiation on aquatic organisms. Adv. Radiat. Biol. 7:103-152.

Boyce, N.P. and S.B. Yamada (1977) Effects of a parasite Eubothrium salvelini (Cestoda: Pseudophyllidea), on the resistance of juvenile sockeye salmon, Oncorhynchus nerka, to zinc. J. Fish. Res. Board Can. 34:706-709.

Bremner, I. (1974) Heavy metal toxicities. Q. Rev. Biophys. 7:75-124.

Brett, J.R. (1958) Implications and assessments of environmental stress. Pages 69-83, The Investigations of Fish Power Problems, edited by P.A. Larkin. Vancouver: University of British Columbia Press.

Brodie, B.B., W.D. Reid, A.D. Cho, G. Sipes, G. Krishna, and J.R. Gillette (1971) Possible mechanism of liver necrosis caused by aromatic organic compounds. Proc. Natl. Acad. Sci. U.S.A. 68:160-164.

Brown, B.E. and R.C. Newell (1972) The effect of copper and zinc on the metabolism of the mussel Mytilus edulis. Mar. Biol. 16:108-118.

Brown, D.A. (1977) Increases of Cd and the Cd:Zn ratio in the high molecular weight protein pool from apparently normal liver of tumor-bearing flounders (Parophrys vetulus). Mar. Biol. 44:203-209.

Brown, D.A. (1978) Decreases of copper and zinc in pretumorous and posttumorous livers of mice exposed to diethylnitrosamine, with and without cadmium or zinc supplementation. Pages 172-196, Toxicology of Trace Metals: Metallothionein Production and Carcinogenesis, edited by D.A. Brown. Ph.D. Dissertation. University of British Columbia. Canadian Thesis Division, National Library of Canada, Ottawa.

Brown, D.A. and A.Y. Chan (1978) Decreases of copper and zinc in the high molecular weight protein pool from livers of mice exposed to diethylnitrosamine, with and without cadmium or zinc supplementation. Pages 144-171, Toxicology of Trace Metals: Metallothionein Production and Carcinogenesis, edited by D.A. Brown. Ph.D. Dissertation. University of British Columbia. Canadian Thesis Division, National Library of Canada, Ottawa.

Brown, D.A. and K.W. Chatel (1978a) Interactions between cadmium and zinc in cytoplasm of duck liver and kidney. Chem.-Biol. Interactions 22:271-279.

Brown, D.A. and K.W. Chatel (1978b) The effect of cadmium exposure, with or without diethylnitrosamine, on the cytoplasmic levels and distribution of cadmium, copper and zinc. Pages 197-237, Toxicology of Trace Metals: Metallothionein Production and Carcinogenesis, edited by D.A. Brown. Ph.D. Dissertation. University of British Columbia. Canadian Thesis Division, National Library of Canada, Ottawa.

Brown, D.A. and K.W. Chatel (1978c) Relationship between survival of mussels exposed to Cd, Cu and Zn, and the cytoplasmic distribution of these metals. Pages 42-58, Toxicology of Trace Metals: Metallothionein Production and Carcinogenesis, edited by D.A. Brown. Ph.D. Dissertation. University of British Columbia. Canadian Thesis Division, National Library of Canada, Ottawa.

Brown, D.A. and B. Knight (1978) Increases of Cd and the Cd:Zn ratio in the high molecular weight protein pool from apparently normal kidney of terminal human cancer patients. Pages 132-143, Toxicology of Trace Metals: Metallothionein Production and Carcinogenesis, edited by D.A. Brown. Ph.D. Dissertation. University of British Columbia. Canadian Thesis Division, National Library of Canada, Ottawa.

Brown, D.A. and T.R. Parsons (1978) Relationship between cytoplasmic distribution of mercury and toxic effects to zooplankton and chum salmon (Oncorhynchus keta) exposed to mercury in a controlled ecosystem. J. Fish. Res. Board Can. 35:880-884.

Brown, D.A., C.A. Bawden, K.W. Chatel, and T.R. Parsons (1977a) The wildlife community of Iona Island jetty,

Vancouver, B.C., and heavy-metal pollution effects.
Environ. Conserv. 4:213-216.

Brown, R.S., R.E. Wolke, S.B. Saila, and C.W. Brown (1977b)
Prevalence of neoplasia in 10 New England populations of
the soft-shell clam (Mya arenaria). Ann. N.Y. Acad. Sci.
298:522-534.

Bubel, A., M.N. Moore, and D.M. Lowe (1978) Cellular
responses to shell damage in Mytilus edulis. J. Exp.
Mar. Biol. Ecol. 30:1-27.

Burns, K.A. (1976) Hydrocarbon metabolism in the intertidal
fiddler crab, Uca pugnax. Mar. Biol. 36:5-11.

Carlson, G.P. (1975) Protection against carbon
tetrachloride-induced hepatotoxicity by pretreatment
with methylmercury hydroxide. Toxicology 4:83-89.

Chen, R.W. and H.E. Ganther (1975) Some properties of a
unique cadmium-binding moiety in the soluable fraction
of rat testes. Environ. Physiol. Biochem. 5:235-243.

Ciapparelli, L., D.H. Retief, and L.P. Fatti (1972) The
effect of zinc on 9, 10-dimethyl-1, 2-benzanthracene
(DMBA) induced salivary gland tumors in the albino rat--
a preliminary study. S. Afr. J. Med. Sci. 37:85-90.

Conover, R.J. (1966) Assimilation of organic matter by
zooplankton. Limnol. Oceanogr. 11:338-354.

Davenport, J. (1977) A study of the effects of copper
applied continuously and discontinuously to specimens of
Mytilus edulis (L) exposed to steady and fluctuating
salinity levels. J. Mar. Biol. Assoc. (U.K.) 57:63-74.

Des Voigne, D.M. and A.K. Sparks (1968) The process of wound
healing in the Pacific oyster, Crassostrea gigas. J.
Invertebr. Pathol. 12:53-68.

Dunn, B.P. and H.F. Stich (1976) Release of the carcinogen
benzo(a)pyrene from environmentally contaminated
mussels. Bull. Environ. Contam. Toxicol. 15:398-401.

Elias, H., A. Hennig, and D.E. Schwartz (1971) Stereology:
Applications to biomedical research. Physiol. Rev.
57:158-200.

Ellis, D.V. (1970) Marine Sediment and Associated Biological
Surveys Around the Crofton Mill Outfall. A report
submitted to British Columbia Forest Products, June 16,
1970.

Feng, S.Y. (1965) Heart rate and leucocyte circulation in
Crassostrea virginica (Gmelin). Biol. Bull. 128:198-210.

Flick, D.F., H.F. Kraybill, and J.M. Dimitroff (1971) Toxic
effects of cadmium: A review. Environ. Res. 4:71-85.

Fong, L.Y.Y., A. Sivak, and P.M. Newberne (1978) Zinc
deficiency and methylbenzylnitrosamine-induced

esophageal cancer in rats. J. Natl. Cancer Inst. 61:145-150.

Fong, W.C. (1976) Uptake and retention of Kuwait crude oil and its effects on oxygen uptake by the soft-shell clam, Mya arenaria. J. Fish. Res. Board Can. 33:2774-2780.

Friedberg, F. (1974) Effects of metal binding on protein structure. Q. Rev. Biophys. 7:1-33.

Gabbott, P.A. (1976) Energy metabolism. Pages 293-355, Marine Mussels: Their Ecology and Physiology, edited by B.L. Bayne. Cambridge: Cambridge University Press.

Gabbott, P.A. and A.J.M. Walker (1971) Changes in the condition index and biochemical content of oysters (Ostrea edulis L.) maintained under hatchery conditions. J. Cons. Cons. Int. Explor. Mer 34:99-106.

George, S.G., B.J.S. Pirie, and T.L. Coombs (1976) The kinetics of accumulation and excretion of ferric hydroxide in Mytilus edulis (L) and its distribution in tissues. J. Exp. Mar. Biol. Ecol. 23:78-84.

George, S.G., B.J.S. Pirie, A.R. Cheyne, T.L. Coombs, and P.T. Grant (1978) Detoxication of metals by marine bivalves: An ultrastructural study of the compartmentation of copper and zinc in the oyster Ostrea edulis. Mar. Biol. 45:147-156.

Goldfischer, S. (1965) The localisation of copper in pericanalicular granules (lysosomes) of liver in Wilson's Disease (hepatolenticular degeneration). Am. J. Pathol. 46:977-983.

Goldman, B.E. (1970) The role of certain metals in axon excitability processes. Pages 275-282, Effects of Metals on Cells, Subcellular Elements, and Macromolecules, edited by J. Maniloff, J.R. Coleman, and M.W. Miller. Springfield, Ill.: Charles C. Thomas.

Gould E. (1977) Alteration of enzymes in winter flounder, Pseudopleuronectes americanus, exposed to sublethal amounts of cadmium chloride. Pages 209-224, Physiological Responses of Marine Biota to Pollutants, edited by F.J. Vernberg, A. Calabrese, F.P. Thurberg, and W.B. Vernberg. New York: Academic Press.

Gray, P. (1954) The Microtomist's Formulary and Guide. New York: Blakiston Co., Inc.

Gunn, S.A., T.C. Gould, and W.A.D. Anderson (1963) Cadmium-induced interstitial cell tumors in rats and mice and their prevention by zinc. J. Natl. Cancer Inst. 31:745-753.

Gunn, S.A., T.C. Gould, and W.A.D. Anderson (1964) Effect of zinc on cancerogenesis by cadmium. Proc. Soc. Exp. Biol. Med. 115:653-657.

Harshbarger, J.C. (1977) Role of the registry of tumors in lower animals in the study of environmental carcinogenesis in aquatic animals. Ann. N.Y. Acad. Sci. 298:280-289.

Hodgson, E. (1974) Comparative studies of cytochrome P-450 and its interaction with pesticides. Pages 213-259, Survival in Toxic Environments, edited by M.Q. Khan and J.P. Bederka, Jr. New York: Academic Press.

International Council for Exploration of the Sea (1978) Report of the Subgroup on the Feasibility of Effects Monitoring. Cooperative Research Report No. 74. Charlottenlund, Denmark: ICES.

Iordachescu, D., I.F. Dumitru, and S. Niculescu (1978) Activation by copper ions of mytilidases: Acid proteolytic enzymes obtained from Mytilus galloprovincialis. Comp. Biochem. Physiol. 61B:119-122.

Ivanovici, A.M. (1978) The adenylate energy charge as an ecological index for monitoring environmental stress in estuarine and marine invertebrates. Paper read at INTECOL, Jerusalem, September 1978.

Ivanovici, A.M. and W. J. Wiebe (1978) For a working definition of "stress": A review and critique. Paper read at INTECOL, Jerusalem, September 1978.

Jeffries, H.P. (1972) A stress syndrome in the hard clam, Mercenaria mercenaria. J. Invertebr. Pathol. 20:242-251.

Jerina, D.M. and J.W. Daley (1974) Arene oxides: A new aspect of drug metabolism. Science 185:573-582.

Johnston, R.E., R.C. Schnell, and T.S. Miga (1974) Cadmium potentiation of drug response: Lack of change in brain sensitivity. Toxicol. Appl. Pharmacol. 30:90-95.

Jones, R.M. (1964) McClung's Handbook of Microscopical Technique. 3rd edition. New York: Hafner Publishing Co.

Khan, M.A.Q., A. Kamal, R.J. Wolin, and J. Runnels (1972) In vivo and in vitro epoxidation of aldrin by aquatic food chain organisms. Bull. Environ. Contam. Tox. 8:219-228.

Khan, M.A.Q., R.H. Stanton, and G. Reddy (1974) Detoxification of foreign chemicals by invertebrates. Pages 177-201, Survival in Toxic Environments, edited by M.Q. Khan and J.P. Bederka, Jr. New York: Academic Press.

Kimbrough, R.D., R.E. Linder, and T.B. Gaines (1972) Morphological changes in livers of rats fed polychlorinated biphenyls. Arch. Environ. Health 25:354-364.

Koenig, H. (1969) Lysosomes in the nervous system. Pages 111-162, Lysosomes in Biology and Pathology, Vol. 2,

edited by J.T. Dingle and H.B. Fell. Amsterdam: North
Holland/American Elsevier.

Kohli, K.K., F.A. Siddiqui, and T.A. Vankitasubramanian
(1977) Effect of dieldrin on the stability of lysosomes
in the rat liver. Bull. Environ. Contam. Tox. 18:617-
623.

Lake, S., B.L. Gledhill, and F.L. Harrison (1977) DNA
variability of abalone sperm as a measure of copper
toxicity. In Proc. of the 17th Hanford Biology
Symposium, October 16-18, 1977, Richland, Washington.

Langston, W.J. (1978) Persistence of polychlorinated
biphenyls in marine bivalves. Mar. Biol. 46:35-40.

Langton, R.W. (1975) Synchrony in the digestive diverticular
of Mytilus edulis. J. Mar. Biol. Assoc. (U.K.) 55:221-
229.

Langton, R.W. and P.A. Gabbott (1974) The tidal rhythm of
extracellular digestion and the response to feeding in
Ostrea edulis L. Mar. Biol. 24:181-187.

Lowe, D.M. and M.N. Moore (1978) Cytology and quantitative
cytochemistry of a proliferative atypical hemocytic
condition in Mytilus edulis (Bivalvia, Mollusca). J.
Natl. Cancer Inst. 60:1455-1459.

Lowe, D.M. and M.N. Moore (1979) The cytolochemical
distributions of zinc (Zn II) and iron (Fe III) in the
common mussel, Mytilus edulis, and their relationship
with lysosomes. J. Mar. Biol. Assoc. (U.K.) (In press)

Malins, D.C. (1977) Metabolism of aromatic hydrocarbons in
marine organisms. Ann. N.Y. Acad. Sci. 298:482-496.

Maugh, T.H. (1978a) Chemical carcinogens: The scientific
basis for regulation. Science 201:1200-1205.

Maugh, T.H. (1978b) Chemical carcinogens: How dangerous are
low doses? Science 202:37-41.

Miller, J.A. (1970) Carcinogenesis by chemicals: An
overview. Cancer Res. 30:559-576.

Mix, M.C. (1975) A review of the cellular proliferative
disorders of oysters (Ostrea urida) from Yaquina Bay,
Oregon. J. Invertebr. Pathol. 26:289-298.

Mix, M.C. (1976) A review of the histopathological effects
of ionizing radiation on the Pacific oyster, Crassostrea
gigas. Mar. Fish. Rev. 38:12-15.

Moore, M.N. (1976) Cytochemical demonstration of latency of
lysosomal hydrolases in digestive cells of the common
mussel, Mytilus edulis, and changes induced by thermal
stress. Cell Tissue Res. 175:279-287.

Moore, M.N. (1977) Lysosomal responses to environmental
chemicals in some marine invertebrates. Pages 143-154,

Pollutant Effects on Marine Organisms, edited by C.S.
   Giam. Lexington, Mass.: D.C. Heath and Co.

Moore, M.N. (1979) Cellular responses to polycyclic aromatic
   hydrocarbons and phenobarbital in Mytilus edulis. Mar.
   Environ. Res. (in press).

Moore, M.N. and D.M. Lowe (1977) The cytology and
   cytochemistry of the hemocytes of Mytilus edulis and
   their responses to experimentally injected carbon
   particles. J. Invertebr. Pathol. 29:18-30.

Moore, M.N. and A.R.D. Stebbing (1976) The quantitative
   cytochemical effects of three metal ions on a lysosomal
   hydrolase of a hydroid. J. Mar. Biol. Assoc. (U.K.)
   56:995-1005.

Moore, M.N., D.M. Lowe, and P.E.M. Fieth (1978a) Lysosomal
   responses to experimentally injected anthracene in the
   digestive cells of Mytilus edulis. Mar. Biol. 48:297-
   302.

Moore, M.N., D.M. Lowe, and P.E.M. Fieth (1978b) Responses
   of lysosomes in the digestive cells of the common
   mussel, Mytilus edulis, to sex steroids and cortisol.
   Cell Tissue Res. 188:1-9.

Moore, M.N., D.M. Lowe, and S.L. Moore (1979) Induction of
   lysosomal destablisation in marine bivalves exposed to
   air (in preparation).

Morton, B.S. (1971) The daily rhythm and the tidal rhythm of
   feeding and digestion in Ostrea edulis. Biol. J. Limn.
   Soc. 3:329-342.

Morton, B.S. (1977) The tidal rhythm of feeding and
   digestion in the Pacific oyster, Crassostrea gigas
   (Thunberg). J. Exp. Mar. Biol. Ecol. 26:135-151.

Noel-Lambot, F. (1976) Distribution of cadmium, zinc and
   copper in the mussel Mytilus edulis. Existence of
   cadmium-binding proteins similar to metallothioneins.
   Experientia 32:324-326.

Oesch, F., P. Bentley, and H.R. Glatt (1976) Prevention of
   benzo(a)pyrene-induced mutagenicity by homogeneous
   epoxide hydratase. Int. J. Cancer 18:448-452.

Owen, G. (1966) Digestion. Pages 53-96, Physiology of
   Mollusca, Vol. II, edited by Wilbur and Yonge. New York:
   Academic Press.

Owen, G. (1972) Lysosomes, peroxisomes and bivalves. Sci.
   Prog. 60:299-318.

Pani, P., A. Sanna, M.I. Brigaglia, A. Columbano, and L.
   Congiu (1976) Early investigations on the effect of
   methyl mercuric chloride upon DMN-acute hepatotoxicity.
   Experientia 32:1449-1451.

233

Pauley, G.B. (1969) A critical review of neoplasia and
tumour-like lesions in molluscs. Natl. Cancer Inst.
Monogr. 31:509-539.
Payne, J.F. (1977) Mixed function oxidases in marine
organisms in relation to petroleum hydrocarbon
metabolism and detection. Mar. Poll. Bull. 8:112-116.
Piscator, M. (1964) On cadmium in normal human kidneys
together with a report on the isolation of
methallothionein from livers of cadmium-exposed rabbits.
N. Hyg. Tidskr. 45:76-82.
Porter, H. (1974) The particulate half-cystine-rich copper
protein of newborn liver. Relationship to
metallothionein and subcellular localization in
nonmitochondrial particles possibly representing heavy
lysosomes. Bioch. Biophys. Res. Commun. 56:661-668.
Poswillo, D.E. and B. Cohen (1971) Inhibition of
carcinogenesis by dietary zinc. Nature 231:447-448.
Prohaska, J.R., M. Mowafy, and H.E. Ganther (1977)
Interactions between cadmium, selenium and glutathione
peroxidase in rat testes. Chem.-Biol. Interactions
18:253-265.
Roberts, D. (1976) Mussels and pollution. Pages 67-80,
Marine Mussels: Their Ecology and Physiology, edited by
B.L. Bayne. Cambridge: Cambridge University Press.
Ruddell, C.L. (1971) Elucidation of the nature and function
of the granular oyster amoebocytes through histochemical
studies of normal and traumatised oyster tissue.
Histochemie 26:95-112.
Ruddell, C.L. and D.W. Rains (1975) The relationship between
zinc, copper and basophils of two crassostreid oysters,
C. gigas and C. virginica. Comp. Biochem. Physiol.
51A:585-591.
Schmidt-Nielsen, B., J. Sheline, D.S. Miller, and M.
Deldonno (1977) Effect of methylmercury upon
osmoregulation, cellular volume, and ion regulation in
winter flounder, Pseudopleuronectes americanus. Pages
105-117, Physiological Responses of Marine Biota to
Pollutants, edited by F.J. Vernberg, A. Calabrese, F.P.
Thurberg, and W.B. Vernberg. New York: Academic Press.
Scott, D.M. and C.W. Major (1972) The effect of copper (II)
on survival, respiration, and heart rate in the common
blue mussel Mytilus edulis. Biol. Bull. 143:679-688.
Selye, A. (1950) Stress. Montreal: Acta Inc.
Shimada, T. (1976) Metabolic activation of ($^{14}$C)
polychlorinated biphenyl mixtures by rat liver
microsomes. Bull. Environ. Contam. Toxicol. 16:25-32.

Stainken, D.M. (1975) Preliminary observations on the mode of accumulation of #2 fuel oil by the soft shell clam, Mya arenaria. Pages 463-668, 1975 Conference on Prevention Control of Oil Pollution.

Sternlieb, I. and S. Goldfischer (1976) Heavy metals and lysosomes. Pages 185-200, Lysosomes in Biology and Pathology, edited by J.T. Dingle and R.T. Dean. Amsterdam: North Holland/American Elsevier.

Stewart, H.L. (1977) Enigmas of cancer in relation to neoplasms of aquatic animals. Ann. N.Y. Acad. Sci. 298:305-315.

Talbot, V. and R. J. Magee (1978) Naturally-occurring heavy metal binding proteins in invertebrates. Arch. Environ. Contam. Toxicol. 7:73-81.

Templeton, W.L., F. Harrison, I. Ophel, J. Simpson, N. Cutshall, H. Volchok, and D. Edgington (1978) Artificial radionuclides. In Ocean Pollution Research, Development, and Monitoring Needs: Report on a Workshop at Estes Park, Colorado, July 10-14, 1978, edited by E.D. Goldberg. (In preparation)

Thompson, R.J., N.A. Ratcliffe, and B.L. Bayne (1974) Effects of starvation on structure and function in the digestive gland of the mussel (Mytilus edulis L.). J. Mar. Biol. Assoc. (U.K.) 54:699-712.

Thompson, R.J., B.L. Bayne, M.N. Moore, and T.M. Carefoot (1978) Haemolymph volume changes in biochemical composition of the blood, and cytochemical responses of the digestive cells in Mytilus californianus Conrad, induced by nutritional, thermal and exposure stress. J. Comp. Physiol. 127:287-298.

Thrower, S.J. and I.J. Eustace (1973) Heavy metals in Tasmanian oysters. Aust. Fish. 32:7-10.

Vallee, B.L. (1976) Zinc biochemistry: A perspective. Trends Biochem. 1:88-91.

Vernberg, W.B. and F.J. Vernberg (1963) Influence of parasitism on thermal resistance of mud-flat snail, Nassarius obsoleta say. J. Exp. Parasitol. 14(3):330-332.

Wada, K. (1978) Chromosome karyotypes of three bivalves: The oysters Isognomon alatus and Pinctada umbricata, and the bay scallop, Agropecten irradians. Biol. Bull. 155:235-245.

Walne, P.R. (1970) The seasonal variation of meat and glycogen content of seven populations of oysters Ostrea edulis, and a review of the literature. Fish. Invest. Lond. Ser. II 26:1-35.

Webb, M. (1972) Binding of cadmium ions by rat liver and
    kidney. Biochem. Pharmacol. 21:2751-2765.
White, A., P. Handler, and E.L. Smith (1968) Principles of
    Biochemistry. 4th edition. New York: McGraw-Hill Book
    Co.
Whittle, K.J., R. Hardy, A.V. Holden, R. Johnston, and R.J.
    Pentreath (1977) Occurrence and fate of organic and
    inorganic contaminants in marine animals. Ann. N.Y.
    Acad. Sci. 298:47-79.
Widdows, J., B.L. Bayne, D.R. Livingstone, R.I.E. Newell,
    and P. Donkin (1979a) Physiological and biochemical
    responses of bivalve molluscs to exposure to air. Comp.
    Biochem. Physiol. 624:301-308.
Widdows, J., D. Phelps, and W. Gallway (1979b) Measurement
    of physiological condition of mussels transplanted along
    a pollution gradient in Narragansett Bay (In press).
Winberg, G.G. (1960) Rate of metabolism and food
    requirements of fishes. Fish. Res. Board Canada,
    Translation Series No. 194.
Winge, D.R., R. Premakumar, and K.V. Rajagopalan (1975)
    Metal-induced formation of metallothionein in rat liver.
    Arch. Biochem. Biophys. 170:242-252.
Yevich, P.P. and C.A. Barszcz (1976) Gonadal and
    hemotopoietic neoplasms in Mya arenaria. Mar. Fish. Rev.
    38:42-43.
Yevich, P.P. and C.A. Barszcz (1977) Neoplasia in soft-shell
    clams (Mya arenaria) collected from oil impacted sites.
    Ann. N.Y. Acad. Sci. 298:409-426.
Yoshida, T., Y. Ito, and Y. Suzuki (1976) Inhibition of
    hepatic drug metabolizing enzyme by cadmium in mice.
    Bull. Environ. Contam. Toxicol. 15:402-405.

# 7

# MONITORING STRATEGIES FOR
# THE PROTECTION OF THE COASTAL ZONE

ROBERT RISEBROUGH (Chairman), University of California,
    Berkeley, California
VICTOR ANDERLINI, Kuwait Institute for Scientific Research,
    Kuwait
MICHAEL BERNHARD, Laboratorio per lo Studio della,
    Lor Spezia, Italy
DEREK V. ELLIS, University of Victoria, British Columbia,
    Canada
G. POLYKARPOV, International Laboratory of Marine
    Radioactivity, Monaco
WILLAM ROBERTSON IV, The Andrew W. Mellon Foundation,
    New York, New York
ERIC SCHNEIDER, Environmental Protection Agency,
    Narragansett, Rhode Island
GRAHAM TOPPING, Marine Laboratory, Aberdeen, Scotland

## ANALYTICAL CAPABILITY

Monitoring strategies may be as diverse as the problems or needs to which programs are addressed. But in all there is a need to ensure that analytical techniques and methodologies employed by different laboratories yield equivalent results. The need is common to all laboratories, those that possess relatively sophisticated "state-of-the-art" instrumentation, and those with relatively simple equipment and analytical techniques.

National and international agencies, including the U.S. NBS and the IAEA, have assumed a degree of responsibility for the distribution of samples or standards for the purpose of intercalibration of analytical methodologies and techniques. The environmental monitoring program in Europe and North America undertaken in the late 1960s and the early 1970s by the OECD included the distribution of samples among

the participating laboratories for intercalibration purposes
(Holden 1973). More recently, laboratories participating in
the Coordinated Monitoring Program in the North Atlantic and
adjacent areas of the North Sea and the Baltic under the
auspices of the ICES have analyzed identical aliquots of
samples distributed by a coordinating laboratory. Most
samples distributed to date have been analyzed for heavy
metals and trace elements; the remainder have been analyzed
for synthetic organochlorine compounds. The program is
continuing and expanding and will, in the future, include
the analysis of environmental samples for petroleum and
petroleum-related HCs. Participation in intercalibration
exercises, however, has usually been limited to those
laboratories participating in the respective monitoring
programs; frequently the number of samples has not been
sufficient to make them available for wider distribution.
Future programs of this kind, including efforts devoted to
the coastal zone, might seek to ensure that sufficient
amounts of the sample are available so that aliquots can be
provided to other laboratories that are undertaking related
or similar studies. Provided with the results previously
obtained by other investigators, these laboratories would be
able to assess and review their own progress in the
application and development of analytical methodologies.
Quality of the results would thereby be improved; data
provided by laboratories undertaking related studies could
then be more readily accepted by the environmental
scientific community. Clearly many scientific activities in
addition to those associated with monitoring could benefit
from this approach.

The training of scientists is also a necessary first
step in the implementation of a program designed to assess
the health of the coastal zones and to provide data to
assist in the formulation of long-term policies of
protection and management. This was recognized by the
United Nations Environmental Program in implementing the
Mediterranean Project. The exchange of scientists among
laboratories and the development of laboratories that would
be equipped to meet local as well as regional needs were
among its initial priorities.


PRIORITIES FOR MONITORING PROGRAMS


Analysis of a few samples of mussels or other bivalves
for a small number of recognized pollutants will not, in
itself, provide any assurance that scientists have

determined the quality of local coastal waters. Nor would such analyses necessarily constitute a basis for a rational program for the long-term protection of the coastal zone. Thus, for example, if heavy metals are analyzed, associated research would be required to determine whether levels are elevated because of the activities of people, and whether higher levels might cause an alteration in local coastal food webs and ecosystems detected or identified as pollutants as well as those that are well known and routinely measured.

Before initiating any monitoring effort, it is desirable to evaluate critically, in so far as can be done in advance, the anticipated value of the proposed program. This is particularly necessary if research funds are to be diverted for monitoring purposes. Research funds that would support the study of natural processes are everywhere in short supply. Unless monitoring can be expected to yield data that would assist in the protection of the coastal zone, other research efforts should probably receive priority.

There is no need to repeat past endeavors. In 1976 the U.S. Mussel Watch Program focused on the DDT and PCB groups among the synthetic organic compounds (Goldberg and others 1978). Except for the desirability of determining rates of change in the levels of these pollutants in selected localities, there would seem to be little point in now repeating the 1976 exercise. Other priorities have emerged. Many unidentified compounds appear in mussel extracts that may be important pollutants now or in the future.

There are a number of potentially important environmental problems that could be amenable to examination through the use of a monitoring program using mussels or other bivalves as sentinel organisms. Among these are unidentified synthetic organics, oil and petroleum products, radioactivity, insecticides, and emissions in coastal areas of rapidly industralizing nations.

## Unidentified or Undetected Synthetic Organics

A plan for long-term protection of the coastal zone from organic pollutants requires consideration not only of those pollutants that can now be detected and measured. The unanticipated occurrence of a pollutant that might result in the closure of local fisheries or have other severe economic consequences presents a challenge to monitoring systems. In the United States the economic cost of the unanticipated discovery of toxic pollutants in coastal and estuarine

waters has probably amounted to many millions of dollars,
much higher than the cost of a monitoring program that might
have detected their presence at an earlier stage. The
discharge of Kepone into the James River in Virginia and of
Mirex into the Niagara River between New York State and
Ontario are among the more familiar examples. In addition
to prompting investigations in the vicinity of pesticide
manufacturing plants elsewhere in the world, events of this
sort illustrate the more general need to anticipate and
thereby prevent such incidents in the future.


### Coastal Pollution by Crude Oil and Petroleum Products

Many beaches in coastal zones adjacent to tanker routes
have been heavily fouled by tar. In addition, wastewater
discharge usually contains high amounts of petroleum-related
or petroleum-derived HCs. The State of California's Mussel
Watch Program has shown that in the vicinity of cities,
levels of HC mixtures in mussels are significantly higher
than they are in relatively pristine areas (Risebrough and
others 1979). The findings have indicated that the
chemistry of local coastal waters is significantly changed
by wastewater discharges. Continuing investigations of the
HC content of waters of inhabited areas in the coastal zone
and of waters along tanker routes might usefully employ the
mussel watch concept.

Methodologies for the measurement of petroleum and
petroleum-related HCs have not yet received as wide an
application as have methodologies for determination of
chlorinated HCs, and fewer laboratories have developed the
capability to undertake these measurements. However, even
relatively crude applications of existing techniques permit
a determination of the relative abundance of petroleum and
of petroleum-derived material in mussels, oysters, and other
bivalves.


### Nuclear Power in the Coastal Zone

As long as the construction and operation of fission-
powered nuclear plants continues, deliberate or inadvertent
leaks of radioactive nuclides can be anticipated. Transport
of radioactive wastes by sea to reprocessing plants provides
another potential pathway of entry of radioactive materials
into coastal environments. The initial survey of artificial
radionuclides in mussels of United States coastal localities

pointed to a number of questions to be resolved in future
research and monitoring programs (Goldberg and others 1978).
The mussel watch concept would appear to be particularly
appropriate in monitoring for leaks in the vicinity of power
plants and for local contamination of food webs if
radioactive wastes should be spilled at sea.

## Insecticide Use in Tropical Countries

Although DDT use in some countries has largely ended,
its use continues in the tropics.  The use of other
relatively persistent pesticides also continues on a
widespread scale, frequently with little regulation.
Problems associated with patterns of pesticide use in
tropical countries might thus be expected, including
contamination of estuaries that are the site of shrimp
fisheries or of aquaculture projects, and harm to local
wildlife, particulary of fish-eating and raptorial birds.
An appropriate strategy in studying this kind of problem,
where contamination by one or more pesticides is suspected
in a local coastal or estuarine environment, might be a
broad survey using a number of indicator species and perhaps
sediments as well.  There is already considerable
information about organochlorine residues in the eggs of
fish-eating birds, but mussels or other bivalves might be
appropriate sentinel organisms for documenting concentration
gradients in the vicinity of a river delta or in coastal
waters near suspected sources of pollution.

## Rapidly Industrializing Coastal States

In most nations that are industrializing rapidly, the
emphasis has not been on pollution control but rather on
development.  With time, however, pollution control becomes
more important, just as it has become in those states now
considered "developed."  The broad survey approach, such as
that first used in the Mussel Watch Programs of the U.S.
Environmental Protection Agency and the State of California,
might yield useful results in monitoring the effects of
increasing industrialization.  The distribution of known
pollutants could be mapped, and hot spots detected.

## PROJECT DESIGN AND IMPLEMENTATION

Fundamental to the concept of mussel watch has been the assumption that mussels and other bivalves are "indicators" of the levels of various substances in the ambient water system and that pollutant levels in mussels might be expected to provide a measure of pollutants in the environment over a large period of time. To date these assumptions have been only imperfectly tested and the extent of their validity has not been established. It has been pointed out in previous chapters, particularly in Chapter 3, that many variables, both external and internal to the organism, may affect the level of uptake of pollutants. Thus, in the State of California's mussel watch program, samples collected within about 100 meters of each other disclosed statistically significant differences in concentrations of cadmium, chromium, copper, zinc, mercury, lead, nickel, and manganese in Mytilus californianus. In the U.S. Mussel Watch Program, no differences in the levels of copper, zinc, and nickel were found between mussels from San Francisco Bay and mussels from adjacent coastal regions (Goldberg and others 1978), although analyses of seawater from San Francisco Bay and offshore indicated higher levels in San Francisco Bay that were associated with anthropogenic inputs (Eaton 1979).

Since many different variables may influence the levels of pollutants accumulated by mussels, observed differences in the levels may not be due to the variables of interest but to other unknown or uncharacterized factors. Such considerations have prompted work toward the development of an "artificial mussel." Polyurethane foam is one potential candidate: it has been used in the extraction of trace organics from seawater and of metal ions from fresh water. The notion is that polyurethane foam suspended in water will behave like an "ideal" mussel in physical and chemical terms, accumulating organics and possibly certain other pollutants from ambient seawater until an equilibrium state is reached for each compound. If developed, such a system would eliminate many of the variables, including physiological condition, size, sex, and reproductive state, that may affect uptake or partitioning in mussels. Moreover, artificial mussels could be used in areas where living mussels do not always occur, particularly in the vicinity of wastewater discharges in harbors and estuaries. But even an artificial mussel system could not eliminate all extraneous variables. Levels of dissolved organic substances in water might profoundly affect partitioning

between the artifical mussels and seawater, just as they
might with living mussels.

The concept of an indicator species is likely to remain
valid, particularly for the majority of the broader surveys.
The validity would be enhanced, however, by a series of
ongoing research projects examining such questions as
determination of mass balances, net fluxes and budgets of
pollutants, their pathways through the environment,
availability, and degradation.

## The Statistics of Sampling

The use of Mytilus and other bivalves as indicators has
in part eliminated a major problem confronting programs that
have used fish. A continuing controversy in using fish has
been whether to analyze samples of or from individual fish,
as has been done in the ICES monitoring programs, or to pool
samples, and if so, how many fish to pool. Pooling reduces
variability so that no estimate of the population variance
can be obtained. On the other hand, population variance is
not the component of variance that is of primary interest;
of more import is the variance associated with space or
time. The analysis of individual mussels has usually not
been feasible largely because a single mussel is too small
to provide enough material for the analysis, and a
controversy over whether extensive analysis of individual
organisms is worth the cost and effort has never arisen in
the mussel watch program.

Nevertheless, variation among individuals does exist as
a result of many different factors. How many individuals
should be pooled for analysis? In approaching this
question, analytical variance must also be considered. The
analytical variance is usually relatively small in the
examination of trace elements. These differences in
replicate analyses are usually less than differences in the
levels recorded among different samples. The same is not
true, however, for the analysis for petroleum, particularly
if the differences among samples is less than a factor of
two. Analytical variability may be comparatively high,
decreasing the ability to detect differences associated with
time or space. The question of how many individual mussels
should be pooled is perhaps best resolved by obtaining
estimates of the variation among individual mussels through
series of analyses of individuals. A simple calculation
then permits a determination of the number of mussels in a
pooled sample that would reduce the sample variance to a

desired level relative to the analytical variance. A significant component of the total variance is thereby eliminated. Many of the remaining variables, such as temperature, levels of dissolved organics in the water, and food supply, can hardly be eliminated in any such sampling strategy. If, however, the sampling site, whether a rock or a pier, remains constant, if the pooling is adequate to reduce sample variance associated with variation among individual mussels, and if the analytical variance is known, the remaining components of variance can be quantified as the residual variance in a regression with time.

## REFERENCES

Eaton, P. (1979) Observations on the geochemistry of soluble copper, iron, nickel and zinc in the San Francisco Bay estuary. Environ. Sci. Technol. 13:425-432.

Goldberg, E.D., V.T. Bowen, J.W. Farrington, G. Harvey, J.H Martin, P.L. Parker, R.W. Risebrough, W. Robertson, E. Schneider, and E. Gamble (1978) The mussel watch. Environ. Conserv. 5(2):101-125.

Holden, A.U. (1973) International cooperative study of organochlorine and mercury residues in wildlife, 1969-1971. Pestic. Monit. J. 7:37-52.

Risebrough, R.W., B.W. de Lappe, E.F. Letterman, J.L. Lane, M. Firestone-Gillis, A.M. Springer, and W. Walker (1979) California Mussel Watch: 1977-1978. Volume III, Organic Pollutants: Mussels, Mytilus californianus, Mytilus edulis Along the California Coast. Draft Report. Water Quality Monitoring Report 79-22. Sacramento, Calif.: State Water Resources Control Board.

PARTICIPANTS

INTERNATIONAL MUSSEL WATCH WORKSHOP


J. ALBAIGES, International Congress on Analytical Techniques
in Environmental Chemistry, Av. M. Cristine, Palau n. 1,
Barcelona 4 SPAIN

VICTOR ANDERLINI, Kuwait Institute for Scientific Research,
P.O. Box 12009, KUWAIT

TURGUT BALKAS, Middle East Technical University, Marine
Science Department, P.U. 28, Erdemli, Icel TURKEY

ANTONIO BALLESTER, Instituto de Investigaciones Pesqueras,
Paseo Nacional s/n, Barcelona 3 SPAIN

MARGUERITA BARROS, Labratorio de Farmacologia, Quinta do
Marques, Oeiras PORTUGAL

BRIAN L. BAYNE, National Environmental Research Council,
Institute for Marine Environmental Research, Prospect
Place, The Hoe, Plymouth PL1 3DH UNITED KINGDOM

MICHAEL BERNHARD, Laboratorio per lo Studio della,
Contaminazione Radioattiva del Mare, Fiascherino - La
Spezia, ITALY

KATHE K. BERTINE, San Diego State University, Department of
Geology, San Diego, California 92057 USA

ELAINE BLANC, SCOPE Secretariat, 51, Blvd. de Montmorency,
Paris 75016 FRANCE

CHARLES R. BOYDEN, Portobello Marine Laboratory, P.O. Box 8,
Portobello, Dunesin, NEW ZEALAND

MARKO BRANICA, Zugreb Laboratories, Ruder Boskovic
Institute, Center for Marine Research, Zagreb YUGOSLAVIA

DAVID A. BROWN, Institute of Oceanography, University of
British Columbia, 2075 Westbrook Mall, Vancouver,
British Columbia, CANADA

KATHRYN A. BURNS, Marine Chemistry Unit, 5 Parliament Place,
Box 41, East Melbourne, Victoria, 3002 AUSTRALIA

PHILIP A. BUTLER, Environmental Research Laboratory,
Environmental Protection Agency, Gulf Breeze, Florida
32561 USA

DANIEL COSSA, Institut National de la Recherche
    Scientifique, INRS-Oceanologie, 310 Ave. de Ursulines,
    Rimouski, Quebec, G5L 3A1 CANADA
DAVID H. DALE, The Papua New Guinea University of
    Technology, P.O. Box 793, Lae Papua NEW GUINEA
RICHARD DOWD, Environmental Protection Agency, 401 M St.
    S.W., 1129 W. Tower, Washington, D.C. 20460 USA
BRUCE P. DUNN, Cancer Research Center, The University of
    British Columbia, Vancouver, British Columbia V6T 1WT
    CANADA
EGBERT K. DUURSMA, Delta Institute for Hydrobiological
    Research, Royal Netherlands Academy of Arts and
    Sciences, Vierstraat 28 4401 EA, Yerseke 806 DFL
    NETHERLANDS
PETER EATON, Environmental Protection Service, Halifax,
    Nova Scotia, B3J 3E4 CANADA
DEREK V. ELLIS, Department of Biology, University of
    Victoria, P.O. Box 1700, Victoria, British Columbia,
    V8W 2Y2 CANADA
RAFAEL ESTABLIER, Instituto Investigaciones Pesqueras,
    Puerto Pesquero, Cadiz SPAIN
JOHN W. FARRINGTON, Woods Hole Oceanographic Institution,
    Woods Hole, Massachusetts 02543 USA
SCOTT FOWLER, International Laboratory of Marine
    Radioactivity, International Atomic Energy Agency,
    Principality of MONACO
R. FUKAI, International Laboratory of Marine Radioactivity,
    Oceanographic Museum, Principality of MONACO
EDWARD D. GOLDBERG, Scripps Institution of Oceanography,
    La Jolla, California 92057 USA
FLORENCE HARRISON, Lawrence Livermore Laboratories, P.O.
    Box 808, Livermore, California 94550 USA
GEORGE HARVEY, U.S. Department of Commerce, Atlantic
    Oceanographic and Meteorological Laboratories, Ocean
    Chemistry Laboratory, 15 Rickenbacker Causeway, Miami,
    Florida 33149 USA
ALAN V. HOLDEN, Department of Agriculture and Fisheries for
    Scotland, Freshwater Fisheries Laboratory, Faskally,
    Pitlochry, Perthshire, SCOTLAND
JANIS HORWITZ, Environmental Studies Board, National
    Research Council, 2101 Constitution Avenue, Washington,
    D.C. 20418 USA
TSU-CHANG HUNG, Institute of Oceanography, National Taiwan
    University, Taipei TAIWAN
RAPHAEL KASPER, Environmental Studies Board, National
    Research Council, 2101 Constitution Avenue, Washington,
    D.C. 20418 USA

JOHN L. LASETER, Center for Bioorganic Studies, University
of New Orleans, New Orleans, Louisiana 70122 USA

MICHEL MARCHAND-STAVRE, CNEXO, Centre Oceanologique de
Bretange, Boite Postale 337, 29.273 Brest Cedex, FRANCE

G. BALUJA MARCOS, Institute of Organic Chemistry, Juan de la
Cierva 3, Madrid 6 SPAIN

JOHN MARTIN, Moss Landing Marine Laboratory, P.O. Box 225,
Moss Landing, California 95039 USA

MICHAEL J. ORREN, Department of Geochemistry, University of
Capetown, Private Bag, Rodebosch 7700, Capetown SOUTH
AFRICA

PATRICK L. PARKER, Marine Sciences Laboratory, University of
Texas, Port Aransas, Texas USA

JEANNIE PETERSON, The Royal Swedish Academy of Sciences,
KVA, Fack, S-104 05 Stockholm, SWEDEN

DONALD K. PHELPS, Environmental Research Laboratory,
Environmental Protection Agency, South Ferry Road,
Narragansett, Rhode Island 02882 USA

DAVID J.H. PHILLIPS, Fisheries Research Station, 100A Shek
Pai Wan Road, Aberdeen, Hong Kong

G. POLYKARPOV, International Laboratory of Marine
Radioactivity, Oceanographic Museum, Principality of
MONACO

JOHN E. PORTMANN, Ministry of Agriculture, Fisheries and
Food, Fisheries Laboratory, Remembrance Avenue, Burnham
on Crouch, Essex SMO 8HA UNITED KINGDOM

ROBERT RISEBROUGH, Bodega Marine Laboratory, University of
California, 2711 Piedmont Avenue, Berkeley, California
94705 USA

WILLIAM ROBERTSON IV, The Andrew W. Mellon Foundation,
140 E. 62nd St., New York, New York 10021 USA

JOAQUIN ROS, Instituto Espanol de Oceanografia, Laboratorio
Del Mar Menor, Apartado Correos, 22, San Pedro del
Pinatar, SPAIN

ILKAY SALIHOGLU, Marine Science Department, Middle East
Technical University, P.K. 28, Erdemli, Icel TURKEY

THEODORE M. SCHAD, Commission on Natural Resources, National
Research Council, 2101 Constitution Avenue, Washington,
D.C. 20418 USA

ERIC SCHNEIDER, Environmental Protection Agency,
Environmental Research Laboratory, South Ferry Road,
Narragansett, Rhode Island 02882 USA

WALTER TELESETSKY, Program Integration Office, National
Oceanic & Atmospheric Administration, 6010 Executive
Boulevard, Rockville, Maryland 20852 USA

GRAHAM TOPPING, Marine Laboratory, P.O. Box 101 Victoria
Road, Aberdeen AB 9 8DB, SCOTLAND

JACK F. UTHE, Research and Development Directorate, Halifax
    Laboratory, P.O. Box 550, Halifax, Nova Scotia B3J 2R3,
    CANADA
GREGORIO VARELA, Facultad de Farmacia, Universidad Central
    Ciudad, Gitania 3 Madrid SPAIN
DAVID L. WEBBER, Department of Botany and Microbiology,
    University College of Swansea, Swansea, Wales
    UNITED KINGDOM
STEVEN WISE, National Bureau of Standards, Center for
    Analytical Chemistry, U.S. Department of Commerce,
    Washington, D.C. 20234 USA
PAUL D. YEVITCH, Environmental Protection Agency,
    Environmental Research Laboratory, South Ferry Road,
    Narragansett, Rhode Island 02882 USA
DAVID YOUNG, South California Coastal Water Research
    Project, 1500 East Imperial Highway, El Segundo,
    California 90245 USA